# MOON

past, present & future

Ben Moore

Copyright © 2019 Ben Moore
All rights reserved for the English text.
Foreign language rights © Kein & Aber AG Zurich-Berlin
Published in German by Kein & Aber in 2019: "Mond: Eine Biografie"
www.benmoore.ch

For Katharina

# Contents

1. Lunar Dreams ................................................................. 1
2. Magnificent Desolation ................................................. 23
3. The Apollo Legacy ........................................................ 51
4. The Origin of the Moon ............................................... 81
5. The First Astronomer ................................................. 100
6. Our Ominous Moon .................................................... 118
7. Illusions of Light and Shadow ................................... 138
8. Mediterranean Musings .............................................. 157
9. Of Astrologers and Astronomers ............................... 177
10. Observing the New World .......................................... 195
11. Lord of the Tides ........................................................ 221
12. Fatal Attraction ........................................................... 244
13. Life in the Moonlight .................................................. 266
14. The Future ................................................................... 289

# Foreword

Fifty years ago, in 1969, the first humans walked on the surface of the Moon, one of the greatest accomplishments of mankind. Now, there is a new space race to the Moon. In the last few years, all of the world's major space agencies as well as several private space companies have announced their goals of establishing a lunar settlement on the Moon by 2030 – the future of lunar exploration is bright.

This magnificent illuminated rock in space still holds many secrets, from its effects and influence on life to our understanding of its origins. We have come a long way in our discovery of the Moon, yet there is still a lot to learn. Over the past fifteen years my research interests have slowly moved from cosmology to planetary origins. I found it frustrating that a compelling theory of the Moon's origin had not been achieved – now we are trying to understand the formation of our Moon using one of the world's largest supercomputers.

As a young theoretical astrophysicist, I began to use powerful computers to simulate our universe, to try to understand the origin of galaxies and the nature of dark matter. But I was also rather envious of my observational colleagues who kept disappearing to spend days and nights on spectacular mountain-top observatories in exotic places. Together with a colleague, we wrote a proposal to observe distant galaxies through a powerful telescope with the grand aim of measuring their distances and motions. We were surprised to obtain two weeks of telescope access on a modest two metre sized instrument, and excitedly we prepared for our observing program. But upon arrival it was not what I expected.

A modern observatory is a high-tech laboratory. The operator sits in a control room away from the actual instrument so that the temperature changes in the telescope dome are minimised. Astronomers never look into the lens of a telescope. Coordinates are entered into a computer and the telescope automatically moves into the correct place and begins to track the object as it moves across the night sky. The telescope collects as many photons as possible and focuses them onto a detector like that in a digital camera, but much larger and more sensitive. The detector is immersed in a bath of liquid nitrogen to reduce thermal noise, and it counts photons literally one at a time, recording their positions and energies. In the control room, the operator can see the images of distant galaxies, invisible to the naked eye, slowly appear on the computer screen.

Staying awake all night was difficult, especially because we could not even open the telescope dome due to the incessant rain that began the day we arrived. After an exhausting week of sleepless nights, the clouds finally began to clear. The conditions were still too poor to allow us to take useful data, so we tried to settle a debate we had about the dark part of the Moon – I thought we could see a dim glow of light beyond the crescent Moon but my colleague thought it was simply an illusion. We decided to see for ourselves by using our telescope to look at the Moon.

When we pointed the telescope at the dark part of the Moon, there was indeed a dim glow of light and we saw in exquisite detail the sheer mountainous walls that rose five kilometres high above a deep impact crater. Suddenly, the sunlit crescent came into view and the number of photons landing on the detector caused it to instantaneously boil the liquid nitrogen. The computer screen blanked out as the system overheated and died, and the observatory was filled with the vapours of liquid nitrogen.

It turns out that the Moon is very bright. Much, much brighter than the faraway things which the telescope was designed to observe – a thousand trillion times brighter in fact. That was the end of observing for the night and from then on, I decided to stick to theoretical astrophysics. At this time, the internet was in its infancy and the vast amounts of information we are used to having at our fingertips today was not readily available. Little did we know that the origin of the dim light on the dark part of the Moon had been sketched and identified as Earthshine by Leonardo da Vinci over 500 years ago.

Astronomy began by observing our Moon. Today, scientists are still making exciting discoveries about our celestial neighbour. From its origins, to the effects of the Moon on our planet and on life, this is a story I have long wanted to tell. In this lunar biography I will describe the history of our Moon, the latest research findings, and what the future holds for our magnificent celestial neighbour.

Since mankind first began to wonder about the cosmos we have dreamed of travelling to the Moon. From folklore and myths to the first science fiction, stories about our Moon inspired those scientists who turned these dreams into reality.

Before we learn of the remarkable achievement made fifty years ago of standing on the Moon, let us begin this journey of discovery with some of the stories that emboldened this endeavour.

---

Ben Moore, Davos, May 2019

# 1. Lunar Dreams

One of my most vivid memories as a child was when my father took me outside to show me the Moon. He told me that there were people up there, right now, walking on its surface. I thought I could see with my eyes the orbiting module that was circling the Moon, waiting for the astronauts to return them to Earth. Of course, that would have been impossible, but not to the eyes and imagination of a child. It was the winter of 1972, and it was the final manned mission to the Moon, Apollo 17. That night, Eugene Cernan made the last footprint on its surface as he stepped back into the lunar module – he later described his experience as being in a science fiction world. I was just six years old and I could not appreciate the enormous achievement of those astronauts, but it left a big impression.

Astronomy began with the aim of keeping track of time and predicting the omens that were associated with the eclipses and planetary alignments. Without the knowledge we have today, the motions and appearance of the Moon would naturally be associated with great powers – the gods. I can only imagine the thoughts of our distant ancestors when unexpected events in the sky occurred. It is enlightening to retrace those ancient steps along the path towards the discovery of our Moon. Some of them were backwards steps, and those are part of our history too.

Us earthlings have always been inspired by our nearest celestial neighbour. It is an intrinsic part of human nature to dream and use our imaginations. Throughout history, dreams of our Moon have been portrayed in numerous stories, poems, myths and legends. How else could primitive societies make sense of the night's sky, dominated by our shining bright Moon? Before the written record it is difficult to reconstruct many of the ancient ideas and beliefs, but some knowledge has been passed down and preserved through generations in folklore and myths. Stories of our Moon are probably as old as stories themselves.

At some point in history our ancestors began to track the occurrence of the seasons by counting cycles of the phases of the Moon. It may have been for the practical purpose of knowing when crops should be planted and harvested, or for the grand ceremonial reason of when to have a big party. Nearly all primitive societies used the regular cycles of our Moon to keep track of time. But I imagine that astronomy also began just because the night's sky is simply beautiful to observe. The story of the discovery of our Moon begins with 30,000 year old Palaeolithic lunar art and continues through the Neolithic era when monuments were constructed, aligned with the yearly and monthly motions of the Sun and Moon.

The oldest sequence of carved lunar phases originates from the Aurignacian culture during the Upper Palaeolithic. Little is known of the Aurignacians – they moved into Europe from the East, supplanting the indigenous Neanderthals around 40,000 years ago. They carved the oldest known artistic representations of humans and animals, and created the earliest known musical instrument, a bone flute discovered in 2008 at Hohle Fels, a Stone Age cave in southern Germany. Around 30,000 years ago an Aurignacian in the Dordogne, France appears to have carved the lunar cycle and the associated phases of the Moon onto a piece of bone. A similarly decorated piece of mammoth ivory, discovered at the Geißenklösterle cave in Germany, dates from the same era. On one side a carved human-like figure appears in worship. On its edge and back there are a series of notches that appear to coincide with the lunar months. The Ishango bone found at the source of the Nile river dates from between 25,000 and 16,000 years ago. Some scientists think that the markings carved onto the fibula bone of a baboon depict a two month lunar cycle.[1] The famous Lascaux cave paintings in France, drawn by the light of fires deep underground 20,000 years ago, show art that is thought to represent the phases of the Moon, the Pleiades star cluster and the Hyades constellation.[2]

In 2004 a series of twelve specially shaped pits arranged in a fifty-metre arc were discovered in Scotland. The central pit is circular and two metres across to match the full Moon, the outer pits resemble the waxing, waning and crescent Moon. Called Warren Field after its location, it dates from 10,000 years ago and is the oldest known lunar-solar calendar. The pits were thought to contain wooden posts that align on the south east horizon with a prominent topographic point where sunrise occurs on the midwinter solstice. It was a tool to keep track of time and the seasons, connecting the solar year with the lunar cycles. The monument had been maintained and periodically reshaped over the course of several thousand years, perhaps hundreds of times in response to shifting solar

---

[1] "The Roots of Civilisation", 1991, A. Marshack, Pub. Moyer Bell Ltd.
[2] *"Eine Himmelskarte aus der Eiszeit?"* [*A Skychart from the Ice Age?*], 1999, M. Rappenglueck, Peter Lang GmbH

and lunar cycles, until the calendar fell out of use around 4,000 years ago.

With organised civilisation came more ambitious monuments constructed from giant stones and aligned with key astronomical events. And there are thousands of such structures across Europe and Asia that were erected during the late Neolithic period. Famous examples include the stone circles in southern Egypt at Nabta Playa and the Goseck circle in Germany which date from 5000BC. The Mnajdra megalithic temple in Malta, the Egyptian pyramids and Stonehenge in England all date from around 3000BC. Many of these Neolithic monuments are aligned with the important solar events of the year, the equinoxes and solstices, when the Sun is observed to rise or set over prominent sighting stones.

### Heavenly powers

Through the earliest written astronomical records from the 3rd millennium BC in Mesopotamia and China and oral records passed down through the ages in India, independently, emerging cultures across our planet were trying to understand the influence of the Moon on their lives.

One of the most important gods to the Sumerians and Babylonians was the male Moon god Sin, or Nanna, represented by a crescent or a bull. The Moon's ever-renewing cycle was interpreted by the Sumerians as a sign of the Moon-god's inherent power to regenerate himself each month. It was believed that he could bestow this power upon all living creatures. Hence, the Moon-god was a fertility god. An important Sumerian text '*The myth of Enlil and Ninlil*' tells the story of the conception of the Moon god Sin and the lord of the underworld Nergal. Written in the 3rd millennium BC, it is perhaps the oldest preserved creation myth.[3] Many of the Sumerian legends were later incorporated into the bible and the Qur'an.

*'The Contendings of Horus and Seth'* is a mythological story from Ancient Egypt written on papyrus around the end of the 2nd millennium BC. There are many versions of the tale of the battle between the gods for the domination of Earth. Seth was the god of the desert and of violence. Horus was a sky deity, and his right eye was said to be the Sun and his left eye the Moon. During one particularly gruesome struggle, Horus rips off one of Seth's testicles and Seth tears out one of Horus's eyes. Seth's mutilation signifies a loss of virility and strength, sometimes associated with the infertility of the desert. The theft or destruction of the Eye of Horus is equated with the darkening of the Moon in the course of its cycle of phases, or accounting for the mysterious eclipses.[4]

In Greek mythology, Selene was the beautiful goddess of the Moon, daughter of the Titans Hyperion and Theia and sister of the Sun god Helios and the goddess of the dawn Eos. Selene drives her silver Moon chariot across the heavens pulled by winged horses. Part of the Homeric *'Hymn to Selene'* reads: *"The air, unlit before, glows with the light of her golden crown, and her rays beam clear, whensoever bright Selene having bathed her lovely body in the waters of Ocean, and donned her far-gleaming raiment, and yoked her strong-necked, shining team,*

---

[3] "Miscellaneous Babylonian Inscriptions", 1918, G. A. Barton, Yale University Press.

[4] "Egyptian Mythology: A guide to the Gods, Goddesses, and Traditions of Ancient Egypt", 2004, G. Pinch, Oxford University Press, page 82-83.

*drives on her long-maned horses at full speed, at eventime in the mid-month: then her great orbit is full and then her beams shine brightest as she increases. So she is a sure token and a sign to mortal men."*[5]

In some stories the Sun and Moon take the role of gods or mystical beings, in others they are husband and wife or brother and sister. Máni is the personification of the Moon in Norse mythology, thought to be connected to the Northern European notion of the Man in the Moon – that the dark grey patches on the Moon appear like the face of a man. A part of the poem Völuspá reads:

*"The Sun, the sister of the Moon, from the south
Her right hand cast over heaven's rim;
No knowledge she had where her home should be,
The moon knew not what might was his,
The stars knew not where their stations were."*[6]

The gender of the Moon may be a consequence of the early lunar myths and folklore. In Latin, French and Greek the Moon is feminine. The Greek goddess of the Moon, Selene, was also called 'Mene', whereas the masculine name 'Men' implied 'month'. It was also the name of the Phrygian Moon-god 'Men'. Her equivalent in Roman religion and myth was Luna, Latin for Moon.

In contrast, the Teutonic tribes regarded the Sun as a female and the Moon as a male deity. The historian Francis

---

[5] Hesiod, Homeric hymns, Homerica XXXII, New York, G.P. Putnam's Sons, 1922.
[6] "The Poetic Edda", 1923, H.A. Bellows, The American-Scandinavian Foundation.

Palgrave, in his 1831 book 'History of the Anglo-Saxons' says: *"contrary to the mythology of the Greeks and Romans, the Sun was considered by all the Teutons as a female, and the Moon as a male deity. They had an odd notion that if they addressed that power as a Goddess, their wives would be their masters."*[7] The legend of the Man in the Moon is similar across northern folklore, often portraying a thieving man who is given the choice of being banished to the hot Sun or the cold Moon and chooses the Moon.

The etymology of the word Moon has its roots in the use of the Moon in measuring the length of the month. The modern German word, Mond, is derived from Old High German māno which itself is derived from the Proto-Germanic word mēnô. From the word mēnô there were various branches such as the English word Moon, Swedish måne or Icelandic máni. The Proto-Germanic word mēnô is derived from the Proto-Indo-European méh₁n̥s (moon, month), which comes from the root *meh₁- 'to measure', the month being the unit of time measured by the Moon.

The month was divided into seven-day weeks by the Babylonians and Sumerians, and the days were given names of the seven celestial bodies, the Sun, Moon and planets. A 17th century BC tablet reveals part of their creation myth: *"Then Marduk assigned to the moon god control of the night and said: 'Each month your tiara of beauty shall shine on the evening's head. For you shall measure the cycles; six days you shall show horns of light; the seventh your crown shall be finished'."*[8] Seven

---

[7] "A history of the Anglo-Saxons", 1831, F. Palgrave, page 52.

[8] "Sources of the Seven-Day Week", 1939, L. S. Copeland, Popular Astronomy, Vol XLVII, no. 464.

days corresponds to the time it takes for the Moon to transition between each phase: full, waning half, new and waxing half. Because the cycle through the lunar phases is 29 and a half days long, the Babylonians would insert one or two days into the final week of each month. Jewish tradition also observes the same seven-day week. The book of Genesis, which adopts a seven-day account of creation, was likely written around 500BC during the Jewish exile to Babylon.

It is not surprising that every emerging tribe and culture came up with myths and stories about the Moon – they were a way of describing phenomenon that seemed unexplainable. The 13th century Norse poem Vafþrúðnismál tells the origin of the Moon and its purpose to keep track of time:

*"Mundilferi is he who began the Moon,*
*And fathered the flaming Sun;*
*The round of heaven each day they run,*
*To tell the time for men."*[9]

In many tribes and societies lunar myths still play an important cultural role today. For example, the Ngas tribe of Nigeria follow a calendar based on regular observations of the Moon. Their agriculture and social activities are fixed by the lunar calendar. They look for the first crescent Moon to appear on the Eastern horizon every month, and their biggest festival is based on observing the first crescent Moon of their New Year. The week-long festival activities include ritually cleaning their homes and their village, giving gifts, and drinking 'Moon beer'. The young boys have their faces

---

[9] "The Poetic Edda", 1923, H.A. Bellows, The American-Scandinavian Foundation.

painted with the full Moon – the 'sons of the Moon' and they shoot arrows into the sky to kill the old Moon in order for the new crescent Moon to be born. The timing of their shooting the Moon has to be precise – the first crescent Moon must be sighted the next evening. If the timing is wrong, the villagers will fall ill. The Ngas also look at the tilt of the first crescent Moon each month since they believe it determines the strength of the seasonal rains. Closer to the equator, the crescent Moon lies on its side with the tips pointing upwards like a bowl. But the Moon also appears to gently tip and tilt over the course of a month because of our changing perspective from the spinning Earth.

That the Moon influences the life and conditions on Earth is interwoven in a great deal of ancient stories. But is there any truth in these ideas? It turns out that reality is even more remarkable than the myths. The life cycle and activities of many species of life on our planet are strongly connected to the lunar phases. Recent research has also revealed a connection between the Moon and the weather on Earth. Over longer timescales, our Moon could be responsible for causing the glacial periods. But our Moon also stabilises our spinning planet, preventing chaotic climate variations. Without our Moon, life on Earth would certainly be very different, and complex life may not have even evolved.

**From astrology to astronomy**

The connections between the cycles of the Moon, the eclipses and life and death were made by many ancient civilisations, and became an intrinsic part of customs and beliefs right across our planet.

'*The Mahabharata*', a 4th century BC Indian epic, describes the eclipses as due to a battle between gods and demons that continues for eternity. The story goes something like this: The gods wished to create Amrita, an elixir of immortality that is made from Samudra manthana – the churning of the ocean of milk. It is a difficult task so they enlist the Asura demons to help, using a mountain to stir the ocean. The gods promise to share the elixir with the demons, but when the task is done, the god Vishnu takes the form of beautiful woman, enchants the demons and takes the elixir. The demon Rahu then sneaks into the camp of the gods and manages to steal a quick drink of the elixir, but the Sun (Surya) and the Moon (Chandra) spot him and warn Vishnu. Vishnu cuts off Rahu's head, but because the demon drank the amrita his head and body have become immortal. Rahu is angry at the Sun and the Moon for warning Vishnu, so his headless body (Ketu) and bodiless head (Rahu) chase them through the sky for eternity. Every once in a while, Rahu catches up with one of his betrayers and swallows them causing an eclipse, but because he's just a severed head, the Sun or the Moon slips back out through his disconnected neck and becomes visible again.

Thus began the earliest myths that tried to describe natural events such as the phases and eclipses of the Moon. Eventually, the myths and stories turned into religions. The difference being that a myth may be a story related to a cosmic phenomenon, whereas a religion turns a story and phenomenon into a deeper meaning, a cause, purpose and consequence. Many ancient civilisations naturally associated the happenings in the cosmos as grand omens from the gods. How else could something be explained that was not understandable by a human? It all had to be the work of something far superior and all powerful. And as myths were transformed into religions, the interpretation of celestial omens was transformed into astrology.

Around the 6th century BC, ancient Greek philosophers were the first to break free from the idea that gods dictated the cosmos and events on Earth. They figured out the real reasons for the eclipses and the phases of the Moon and even measured the Moon's distance and size. But rational thought ended abruptly with the Roman Empire, and the likes of Ptolemy and Pliny the Elder set scientific reason back a thousand years. The Moon was speculated to influence human behaviour, astrology flourished and the dreams of the Moon turned to nightmares during the mediaeval period. The light from the full Moon was thought to drive people to lunacy – a myth that originated from a fusion of Sumerian and Roman beliefs and the association of the Moon with causing the tides. Because of its ability to move oceans of water, it was widely believed that the Moon must affect all life on Earth.

From ancient times until the 17th century, our Moon was considered to be one of the planets, all contained within crystalline shells that moved them around the Earth together with the Sun. The most important of the planets was the Moon, and it was commonly believed that the Moon was made by some all-powerful creator. Scientific theories for its origin only emerged after the invention of the telescope and after religion had relaxed its rigid grip on free thinking. Descartes' work *Le Monde* was written in 1630, and described his theory of vortices for the origin of the Earth and Moon. Over the past four hundred years there have been many attempts to explain how our magnificent Moon came to be. But none have withstood the test of time and even today, scientists do not agree on the details as to how our Moon formed.

**Imagination and inspiration**

With so little knowledge of what our Moon was really like, we could only speculate. During the early period of ancient Greece, some philosophers thought that the Moon was a gateway into a ring of fire. Others wrote that it must reflect the light of the Sun and be Earthy in character. That led to a great deal of freedom for writers, poets and playwrights to imagine the Moon and create wonderful portrayals of journeys to the Moon to meet the beings that inhabited it. Their writings did not suffer from the inhibitions of science fiction today since so little was known. This allowed a freedom of imagination that today would be considered as fantasy.

Almost two thousand years ago the Roman author Lucian of Samosata wrote *'True History'*, a satirical story in which Lucian and his companions are lifted into the sky by a whirlwind which carries them to the surface of the Moon. Once there, the adventurers are caught up in a political war between the kings of the Moon and the Sun over who should have the right to colonise Venus.

The 10th century Japanese text *'The Tale of the Bamboo Cutter'* is a wonderful short story about a celestial being from the Moon who had been banished to Earth from the palace of the Moon. After taking on a beautiful human form and spending twenty years on Earth, falling in love and capturing the hearts of many, the heroine is taken against her will back to the palace of the Moon while humanity gazes helplessly at her plight.

The invention of the telescope at the beginning of the 17th century led to widespread excitement – it was as if a new world had been discovered, and people were fascinated by the prospect of viewing its surface. This magnificent shining light in the night's sky was studied through ever larger telescopes for over three hundred years. Astronomers speculated as to the presence of rivers and oceans, trees and animals. There were remarkable claims of witnessing life on its surface, but their telescopes could not resolve anything smaller than a few kilometres across. Until the first spacecraft landed intact on the Moon in the 1960's, the nature of its cratered surface was a mystery.

Many imagined that our Moon must host life, and dreamed of visiting its surface. But what could that life be like? Johannes Kepler was amongst the first to study the Moon through a telescope. In 1608, he wrote his own story, 'Somnium', although it was not published until 1634. In his dream, the narrator is carried high above the Earth by a daemon (a benevolent intelligence), travelling in the shadow of an eclipse before falling to the surface of the Moon. Kepler describes what it is like to stand on the near and far side of the Moon, the lengths of night and day and the eternal view of the Sun from the poles. It is as if Kepler was explaining the new Copernican view of the solar system, using his science fiction story as a means of avoiding the wrath of the church.

Francis Godwin wrote '*The Man in the Moone*' at the beginning of the 17th century, possibly even before Kepler's '*Somnium*', although it wasn't published until after his death in 1638. The hero of the story is named Domingo Gonsales, who is carried in a device pulled by a breed of swans that migrate to the Moon each year. Gonsales finds an inhabited world with oceans of water and a race of tall Christian people enjoying a utopian paradise. Godwin also describes the motion of the Moon and Earth but does not completely embrace the then new Copernican worldview – perhaps because he was a bishop of the church.

The French novelist and playwright Cyrano de Bergerac wrote '*L'Autre monde ou les états et empires de la Lune*' (The Other World: Comical history of the states and empires of the Moon) in 1657. It is a satirical novel where the narrator, Cyrano, unsuccessfully tries to reach the Moon with a water powered device. The aim was to reach the Moon and prove that there was a civilisation there that views the Earth as its Moon! Cyrano eventually reaches space but with a flying machine powered by fireworks. Little did he know that what he intended as humour would turn out to be the basic principle of space travel several centuries later. During his journey, Cyrano meets the ghost of Socrates and Domingo Gonsales, the character from Francis Godwin's book. He talks with Gonsales about the futile concept of God.

Rudolph Erich Raspe's '*Baron Munchausen's Narrative of his Marvellous Travels and Campaigns in Russia*' (1785) tells a story in which the lunar visitors are blown to the Moon in a gale. In George Tuckers 1827 novel '*A Voyage to the Moon*', an antigravity material is used to propel voyagers to the Moon.

In other works from this period, springs and balloons and all sorts of contraptions were used to take people to the Moon. The Moon as a utopian paradise filled with life reflected the excitement about this new world that was being studied in ever greater detail through the world's largest telescopes.

Jules Verne's 1865 book *'De la terre a la lune'* describes a space gun to fire a capsule to the Moon – a concept that was later used in the first science fiction movie about the Moon, the spectacular *'Le Voyage dans la Lune'* by Georges Melies in 1902. Jules Verne wrote 'hard' science fiction, as realistic as possible. He was perhaps the first to portray a realistic space capsule and to accurately describe being weightless in space long before anyone had been into space. Since he could not envisage a scientific way of returning his travellers from the Moon's surface, they circled the Moon and returned to Earth.

*A scene from George Melies 'Le Voyage dans la Lune'*

By the time the cosmic voyage began to be taken seriously towards the end of the nineteenth century, scientists had become sceptical about advanced life on the Moon. There was no evidence for an atmosphere, water or weather on the Moon. And astronomers had just made the first measurements of the extreme temperatures on its surface. This was reflected in the science fiction stories of the time – several authors portrayed the Moon as a desolate world where life was extinct, whereas in other stories lunar travellers find the ruins of long dead civilisations. In his 1901 book '*The First Men in the Moon*', H. G. Wells describes an antigravity material that is used to steer a craft to the Moon. The two travellers discover a desolate landscape but find an advanced insectoid society living within the Moon.

**Realising the dream**

Just as the astronomers' studies of the Moon inspired novels, films and science fiction, the fictional stories inspired a new generation of scientists to pursue the dream of travelling to the Moon. Konstantin Tsiolkovsky was born in 1857 in central Russia and was amongst the first to consider the details of the exploration of space using rocket technology. He was a reclusive home-schooled child because of illness and spent much of his time reading books about science and science fiction.

He dreamed of colonising space and was the first true rocket scientist – he derived the 'rocket equation' in 1897 which carries his name today. In 1903 he published his most famous work '*Exploration of outer space by means of rocket devices*', in which he calculated the energy required for space travel and proposed a multi-stage rocket fuelled by liquid oxygen and hydrogen. Tsiolkovksy's rocket equation and multistage design became the basis for all subsequent space flights. In 1911 he wrote in a letter to a friend "*The Earth is the cradle of humanity, but we cannot live forever in a cradle.*" Despite his belief in the diversity of life out there and his advocacy of human spaceflight, he never actually tested his ideas by building a rocket. Several decades later, scientists from the Soviet Union began to turn his ideas and dreams into reality.

In 1893 Tsiolkovsky published a science fiction novel '*On the Moon*' in which he accurately portrays what it would be like to stand on the Moon with the effects of its low gravity. In his 1903 publication on space travel, he refuted Verne's idea of using a cannon for space travel. He calculated that a gun would have to be impossibly long and would subject the travellers to a g-force of over 20,000 which would immediately turn any astronaut into a biological mush. However, he was nevertheless greatly inspired by the story. He states in his epitaph: *"Man will not always stay on Earth; the pursuit of light and space will lead him to timidly penetrate the bounds of the atmosphere and then finally conquer the whole of outer space."*

Tsiolkovsky was not the only person inspired by science fiction to become a rocket scientist. In 1913 the French aircraft designer Esnault Pelterie independently derived and published the rocket equation unaware of the work of Tsiolkovsky. During his 1912 address to the French Physical Society he stated *"Numerous authors made a man traveling from star to star a subject for fiction ... No one has ever thought to seek the physical requirements and the orders of magnitude of the relevant phenomena necessary for the realisation of this idea ... This is the only aim of the present study."* In 1930 he proposed the use of atomic energy to power interplanetary spacecraft.

Another pioneer of rocket science was the German physicist Hermann Oberth (1894-1989). At the age of 11 he read and reread Jules Verne '*From Earth to the Moon*' and '*Around the Moon*' until he had virtually memorised the texts. Fascinated by the idea of travelling into space on a rocket, he constructed his own model rocket at the age of 14. With that experience he also came up with the idea of a multi-stage rocket. In 1922 he wrote his dissertation and what would become a famous book on rocket science: *'Die Rakete zu den Planetenräumen'*. Later, Oberth would be recruited by Wernher von Braun to help create the V-2 rocket during the Second World War. Like many rocket scientists, Oberth was inspired by the ideas of exploring the universe and in 1957 he wrote *"This is the goal – to make available for life every place where life is possible. To make inhabitable all worlds as yet uninhabitable, and all life purposeful."*[10]

Oberth was a scientific consultant on the first realistic science fiction movie, *'Frau im Mond'* (1929). This wonderful silent movie was directed by Fritz Lang and based on the novel of the same name by Thea von Harbou, his wife at the time, who also wrote the classic movie *'Metropolis'*. *'Frau im Mond'* contains many of the discoveries and speculations about our Moon that were made by astronomers during the previous decades. Once you have read the later chapter on early telescopic observations of the Moon, if you watch this silent movie you will see the influences from astronomers at the time.

The American rocket scientist Robert Goddard was also inspired by space exploration as a child when he read H.G. Wells' classic *'The War of the Worlds'* aged 16. During graduation Goddard said, *"It is difficult to say what is impossible, for the dream of yesterday is the hope of today and the reality of tomorrow."* Goddard launched the first liquid fuelled

---

[10] "Man into Space: New projects for Rocket and Space Travel", 1957, H. Oberth, chapter VIII, page 167, translated by G.P.H. de Freville, Harper, New York.

rocket in 1926 achieving an altitude of over two kilometres and a launch speed of over 800 kilometres per hour. Goddard was unaware of the work of Tsiolkovsky and patented a multi stage liquid fuelled rocket in 1914. He made many new innovations such as the use of three axis gyroscope controls to stabilise rockets in flight.

These pioneers of rocket science inspired a new era of realism with many thinking that humans could soon be on the Moon. Popular articles were written that lent further inspiration, such as Arthur C. Clarke's 1939 essay, *'We Can Rocket to the Moon – Now!'* published in the magazine Tales of Wonder. After the Second World War, which saw the fearful V-2 rocket cross the skies, a number of near-future science fiction stories were written about the first Moon landings. These included a series of excellent novels by Robert Heinlein such as his 1950 book *'The man who sold the Moon'*.

In his popular astronomy book *'A guide to the Moon'* (1953), the astronomer Patrick Moore wrote *"Our Earth has been ransacked. Modern man sighs for new worlds to conquer, and the solar system awaits his inspection. He has the ability, and he is gaining the knowledge; and provided that he keeps his senses, he stands on the threshold of the Space Age."* These were wise words, written at a time that must have been rather frightening. The space race between the United States and the Soviet Union was already well underway. Rockets were being developed and tested with the aim of being able to carry nuclear weapons between continents.

By 1957 the Soviet Union had succeeded in launching Sputnik, the first artificial satellite to orbit the Earth. And in

1958 the American Space Agency NASA was founded, and there was already talk about the science that could be accomplished via the exploration of the solar system with the primary destination being the Moon.

Arthur C. Clarke's 1961 science fiction novel '*A Fall of Moondust*' portrays our knowledge of the Moon shortly before the first spacecraft visited its surface. Set in the 21$^{st}$ century, after the Moon has been colonised, it is visited by wealthy tourists who can afford the journey. One of the tourist attractions is to ride across one of the lunar mare in a vehicle similar to a jet ski. The unfortunate space tourists sink beneath the dusty lunar surface, and it's a race against the clock to rescue them.

In his address to the United States Congress in 1975, Clarke said, "*I'm sure we would not have had men on the Moon if it had not been for Wells and Verne and the people who write about this and made people think about it. I'm rather proud of the fact that I know several astronauts who became astronauts through reading my books.*" Arthur C. Clarke also wrote 'hard' science fiction that sticks to scientific plausibility. Indeed, prior to the Moon landings, it was not known what the lunar surface was like, and many scientists and the Apollo astronauts were worried that their landing craft might sink into a deep layer of Moondust. Clarke's story of space tourists is rather fitting since this is now a major driver of the private rocket industry.

I will get to the future of our Moon much later, now it is time to tell the story of how the dreams of these first rocket scientists and science fiction authors were realised. Dreams that ultimately led to humans walking on the Moon's surface and returning evidence that allowed us to decipher its history and origins. Those dreams have never faltered and a new space race has recently begun. In the last few years, several space agencies have not only announced their intentions to return to the Moon, but to build permanently manned Moon-bases by 2030. It is not only nations and scientists that dream of visiting the Moon again. A new generation of billionaires, enabled by private space industries, want to make their own dreams come true.

## 2. Magnificent Desolation

Fifty years ago the first humans set foot on the Moon, watched live by a fifth of the world's population. This was achieved by those who were inspired by the dreams of those who came before, and its legacy has fuelled the dreams of those who came after.

The story of the first Moon landing could begin at many points in human history, but let's begin with rockets. Getting to the Moon needed a vehicle that could counter Earth's gravitational pull and fly through the vacuum of space, safely transporting a life-form that is not designed to exist beyond the confines of Earth's lower atmosphere. And that's just the beginning of the journey.

The first rockets were constructed by the Chinese in the 13th century – these chemically propelled fireworks were used in the war against the Mongols with great success. Fireworks use a form of gunpowder, which is rather different from the rocket fuel used today, but it can be made with materials that can be easily produced – sulphur, charcoal and saltpetre (potassium nitrate) which acts as the oxidiser – the chemical that can rip the electrons from other elements, binding them more tightly and releasing energy in the process.

It was generally thought that little progress had been made until the start of the 20th century. But in 1961 an old manuscript was discovered that revealed early insights on rocket science. It was written in the 16th century by an Austrian military engineer named Conrad Haas and contained a discussion of rocket technology, combining the use of fireworks with weapons. Haas sketched multi-stage rockets and proposed the use of liquid fuel and aerodynamic fins for stability. He may have even anticipated the idea of the modern spaceship with a sketch of a rocket propelled house. Despite developing weapons technology he concluded his book with a humanist view: *"But my advice is for more peace and no war, leaving the rifles calmly in storage, so the bullet is not fired, the gunpowder is not burned or wet, so the prince keeps his money, the arsenal master his life; that is the advice Conrad Haas gives."*

Whilst the primary motivation for the space race to the Moon came from the politics of the cold war, the technology that enabled it stemmed from those scientists who had been inspired by the dreams of travelling to the Moon and beyond. But there was still an awful lot of technology that needed to be developed. No science fiction story before the Moon landings portrayed the full complexity of carrying humans to the Moon and back. Nearly everything had to be designed from scratch.

### From fireworks to rockets

Modern rocket science began, as I mentioned in the last chapter, independently on different continents with Konstantin Tsiolkovsky, Hermann Oberth, Robert Esnault-Pelterie and Robert Goddard. It began at a time when it was thought that the Moon and planets hosted life that was within reach of discovery and was enabled by the scientists that had paved the way with a deeper understanding of astronomy, mechanics, fluid dynamics, chemistry and mathematics.

Tsiolkovsky derived his rocket equation using basic physics – the conservation of momentum. This enabled him to calculate the acceleration and top speed of a rocket that carries a payload of a given mass. Since the rocket engines produce a constant thrust, and as the rocket loses mass by burning its fuel, it slowly accelerates to higher speeds. The heavier the payload, the more fuel is necessary to reach outer space. Tsiolkovsky calculated that most of the mass of the initial rocket had to be fuel and proposed having separate booster stages to save weight since you can jettison the empty fuel containers along the way.

The first rocket scientists envisaged powering them with a mixture of hydrogen and oxygen. This is a fuel adopted by many rockets today, but some use ethanol and oxygen, or petrol and oxygen. The oxygen is the 'oxidiser' since it is one of the most reactive elements known, tightly bonding to most other elements and releasing energy in the process.

The principle of rocket propulsion is easiest to understand if you imagine a common variant of rocket fuel – water. Water is made of two hydrogen atoms bonded to an oxygen atom. If you provide enough energy, the water molecules can be split apart into hydrogen and oxygen. If that process were easy it would be a great source of carbon friendly fuel, but unfortunately it takes a lot of energy to separate those tightly bonded atoms. Conversely, if you take two tanks of hydrogen and oxygen and allow them to mix, the oxygen bonds with the hydrogen releasing energy in the form of rapid kinetic motions of the resulting water molecules.

A rocket engine works by controlling the mixture of hydrogen and oxygen, burning a small amount of fuel and allowing the waste products of the resulting molecules to shoot out of a nozzle at the back, reaching speeds of four kilometres per second. If the mix is not controlled then the resulting chain reaction will result in an explosion.

The amount of fuel needed to reach space depends on the gravitational pull of the planet, the mass of the empty rocket and its payload and the mass of fuel before launch. One litre of water contains enough rocket fuel to overcome Earth's gravitational pull and propel just over one kilogram into space! That's good, because it allows for payload that is a few percent of the mass of the fuel. If Earth were a slightly larger planet with a stronger gravitational pull, reaching the Moon would have required far larger rockets than the giant Saturn V rocket used for the Apollo program.

But there is much more complexity to a rocket than to a simple firework. The rocket has to be stable at speeds much higher than the speed of sound. The liquid fuel needs to be stored at very low temperatures, and this causes ice to form on the outside of the rocket tanks. The fuel mix has to be just right and ignited in a combustion chamber that has to withstand temperatures of several thousand degrees. Rockets need to work in Earth's atmosphere and in the cold vacuum of space. The rockets are not usually piloted and everything has to be automated – gyroscopes are used to track their orientation, and a multitude of sensors provide information to on-board computers that control their motion.

The complexity of a manned space rocket is similar to a jumbo jet. The Airbus A380 has several million individual parts produced by over a thousand different companies. That's not too different from a rocket that takes astronauts to the International Space Station. However, the A380 is built upon generations of previous planes with thousands of test flights, whereas the rockets used to reach the Moon were designed and constructed within five years and were too expensive to test more than a handful of times.

The development of rocket technology began in earnest in Germany during the Second World War. Scientists who had dreamed of space flight were recruited to design ballistic missiles that could travel between countries with a payload of explosive warheads rather than people. In 1942 a prototype of the V-2 rocket became the first man-made object to reach outer space. With a range of over 300 kilometres and supersonic speeds as high as 5700 kilometres per hour, it could carry a destructive payload of more than a ton. It was developed and used to bombard England during 1944-1945.

After the war, the V-2 became the basis of early American and Soviet rocket designs. British, American and Soviet intelligence teams competed to collect Germany's rocket engineers. The Soviets captured 170 German rocket specialists and moved them to a remote location to help develop their own rockets. The Americans took the lead of the V-2 project, Wernher von Braun, and over one hundred of his colleagues and relocated them to White Sands in New Mexico. The cold war had begun in earnest.

Once allies during World War 2, the Americans and the Soviet Union entered a psychological war, a struggle between the ideals of communism and capitalism – a struggle that had been brewing for several decades. The subsequent build-up of military forces by both nations was quite terrifying.

Already by 1945 the Americans had dropped their first nuclear bombs. By 1949 the Soviets had caught up and began testing their own atomic bombs. This was followed by the Americans announcing they would build an even more destructive weapon, the hydrogen fusion bomb that was capable of destroying even the largest of the world's cities.

**The race to space**

Rocket technology needed to be improved so that these heavy weapons could be launched halfway across the planet. The cold war was fought on many fronts, including in space. A secondary motivation for developing rocket technology was the need to deploy spy satellites. The space race was well underway by 1955 when both the USA and the Soviet Union announced they would launch artificial satellites by 1958.

By 1957 the Soviets had developed the R-7 Semyorka rocket which was the first operational intercontinental missile. It was a much improved design based on the V-2 rocket. But as nuclear warheads became smaller, such a heavy lifting rocket was not needed for carrying missiles anymore. Instead it was used as the launch vehicle for the first satellites and astronauts – or cosmonauts as the Soviets named their space travellers. A modified version of the Russian R-7 is still in use today to carry supplies and astronauts to the International Space station.

By the same year the Americans were close behind the Soviets with their Atlas-A rocket, which was developed by the United States Air Force. Concurrently, the United States Navy was developing the Redstone rocket, which more closely followed the V-2 design. This would be advanced to become the heavy lifting rocket, Saturn V, which was eventually used by the Apollo program.

On October 4th 1957 the Soviets launched the first artificial satellite into space. Using a modified R-7 rocket, the payload was Sputnik, a half metre in size and weighing 84 kilograms. Most of its weight was its power supply, used to transmit a repeating series of beeps in the radio wavelengths as it circled the Earth. The launch of Sputnik shocked the Americans. The following month the Soviets launched the dog Laika into space. That shocked the world even more, especially those who thought it was a cruel thing to do.

Things got worse when President Eisenhower brought forward the timetable for the first American satellite launch. On December 6th 1957, the Vanguard rocket lifted one metre above the launchpad at Cape Canaveral and crashed back to Earth in a huge fireball two seconds later, an event that was broadcast live. In the United Nations, the Soviet delegate sarcastically offered the US representative aid *'under the Soviet program of technical assistance to backwards nations'*, humiliating the Americans even further.

In 1958 the Americans successfully launched their first satellite into space. In the same year President Eisenhower created the National Aeronautics and Space Administration (NASA), a federal agency dedicated to non-military space exploration, as well as several additional programs seeking to exploit the military potential of space. With the successful launch of satellites, the potential of exploring space became a reality, and scientists began to dream of utilising the new technology for exploring the solar system.

In 1958 the American scientist Harold Urey visited NASA headquarters to press for a serious scientific study of the Moon. Urey and his scientific advisory teams to NASA advocated that the Moon was a Rosetta stone of planetary origins. They pointed out that the lack of atmosphere and water on the Moon meant that its surface was not subject to erosion, which over geologic time had erased much of the history of the Earth. They made the compelling scientific case that a study of the Moon would reveal an understanding not only of the origin of the Moon, but also of the Earth and the solar system.

Urey was one of the most famous American scientists, having received a Nobel Prize for the discovery of Deuterium. He had many interests and developed the field of cosmo-chemistry as a means of studying meteorites and planetary surfaces, as well as carrying out pioneering experiments on the origin of life. NASA took him seriously and a working group formed shortly after to discuss what experiments could be done via space missions. This motivated NASA to begin its Ranger exploration program, aiming for the Moon.

Still at this time, the Soviets were several steps ahead. The first of the Soviet Luna mission launches took place in 1959, also with the then undisclosed aim of reaching the Moon. Luna 1 malfunctioned and narrowly missed the Moon by a distance of about 6000 kilometres after a journey of nearly 400,000 kilometres. Luna 2 reached the Moon in September 1959 after travelling for 36 hours. Its upper stage was a 390 kilogram satellite that transmitted information back to Earth. It was deliberately crashed into the Moon's surface and became the first man-made object on the Moon. The following month Luna 3 orbited the Moon, took the first photographs of its far side and beamed them back to Earth. A remarkable achievement that revealed for the first time what no human had ever seen.

Aware of the Soviets' lead, in 1958 the United States Air Force were developing a secret project named A119 in which plans were made to detonate a nuclear bomb on the Moon. The existence of the project was revealed only in the year 2000 by the head of the project, but it has never been officially admitted to. Only very recently has evidence surfaced that the Soviets were working on a similar project. Thankfully, these ridiculous shows of force never took place for the fear of the negative publicity surrounding the militarisation of space. Later, the Outer Space Treaty, signed by the Soviet Union and America in 1967, forbid the use of nuclear weapons in space.

In 1960 John F. Kennedy won the presidency of the United States with the slogan *'Let's get this country moving again'*, highlighting the gap with the Soviets on various fronts. Attempting to recover from the humiliation in space technology against the Soviets, President Kennedy consulted with the rocket scientists Jerome Wiesner and Wernher von Braun as to how to catch up with the Soviets. Long term plans of either a space station or a Moon landing were discussed. Von Braun convincingly argued that a Moon landing could be realistically achieved.

The idea of a Moon landing had already been conceived when NASA began a feasibility study. At this time the program was named 'Apollo' after the Greek god by NASA's director of space flight development, Abe Silverstein. He chose the name Apollo to follow the tradition of using names from Greek mythology for rockets and missions, and for the reason that Apollo was one of the most important of the Greek gods. The initial progress on Apollo was slow. Until, in April 1961, the Soviets surprised the world again when they launched Cosmonaut Yuri Gagarin into orbit around the Earth. Two days after the Gagarin flight Kennedy met with the head of NASA, James Webb, to discuss the possibility of landing Americans on the Moon.

President Kennedy unveiled the commitment to execute Project Apollo on 25 May 1961 in a speech on 'Urgent National Needs' announced as a second State of the Union message. He told Congress that the United States faced extraordinary challenges and needed to respond extraordinarily: *"If we are to win the battle that is going on around the world between freedom and tyranny, if we are to win the battle for men's minds, the dramatic achievements in space which occurred in recent weeks should have made clear to us all, as did the Sputnik in 1957, the impact of this adventure on the minds of men everywhere who are attempting to make a determination of which road they should take. ... We go into space because whatever mankind must undertake, free men must fully share."* Then he added: "*I believe this Nation should commit itself to achieving the goal, before this decade is out, of landing a man on the moon and returning him safely to Earth. No single space project in this period will be more impressive to mankind, or more important for the long-range exploration of space; and none will be so difficult or expensive to accomplish.*"

His speech was certainly inspiring and received strong support. The deadline he announced of the end of the decade was an ambitious goal, and it needed a huge amount of money to be realised. But this was approved by congress, and NASA's annual budget was rapidly increased from 500 million dollars a year in 1960 to a high of 5.2 billion dollars in 1965 – over five percent of the federal budget.

The task of getting humans to the Moon and back was enormous. All that existed at the time were rockets that could carry a payload of a satellite to a low Earth orbit. Everything else still needed to be developed, from larger rockets that could carry lunar landers and astronauts to communication, life support systems and a way of returning the astronauts to Earth. In 1960 NASA employed less than 10,000 people. By 1966 this had grown to over 400,000 people working directly or indirectly for the Apollo program led by NASA. Most of this manpower came from outsourcing expertise to companies and universities. Reaching the Moon became a management task, and NASA oversaw the planning, outsourcing procurements, controlling the hardware and getting the funding. Everything had to work just right down to the last screw.

In today's money, the Apollo program cost well over a hundred billion dollars. It involved over 20,000 companies and universities across America. All of this was coordinated by the Apollo program manager, George Mueller. The project was a masterpiece in management and organisation. James E. Webb, the NASA Administrator at the height of the program between 1961 and 1968, stated that Apollo was much more a management exercise than anything else.

At the time Kennedy announced the Apollo program, the Soviets did not confirm or deny that they were pursuing the same goal. Only much later it was disclosed they were also attempting to land cosmonauts on the Moon, until their plans were abandoned shortly after the conclusion of the Apollo program. In 1961 the course of history almost changed when Kennedy attempted to ease the cold war by inviting the Soviets to collaborate on the effort to reach the Moon. The Soviet premier Nikita Khruschchev at first rejected the proposal. However, his son later confessed that he was going to accept Kennedy's offer in 1963, but thought otherwise after Kennedy's assassination since he did not trust his successor Lyndon B. Johnson. That is not surprising since as the Senate majority leader in 1958, Johnson had declared at the Democratic Conference: *"Control of space means control of the world."*

In 1961 NASA's Ranger program began launches with the aim of taking the first close up images of the Moon. Ranger 1 and 2 suffered from launch failures. Ranger 3 reached the Moon but on the wrong trajectory and flew straight past. In 1962 Ranger 4 reached the Moon but crashed on the far side without transmitting any data. In the same year Ranger 5 lost power before reaching the Moon. Success was achieved in 1962 when John Glenn became the first American to orbit the Earth as part of the Mercury mission. In 1964 Ranger 6 reached the Moon and achieved its crash landing, but the camera failed to operate. Finally, a few months later, Ranger 7 achieved its goal of transmitting close up images of the Moon before deliberately impacting onto its surface.

The Soviets achieved many other 'firsts' in the following years, including the first woman in space, the first spacewalk, the first two and three man crewed space craft and the first cosmonauts to fly in normal clothes rather than in space suits from launch. However, with the greatly increased NASA budget, the Americans began to move ahead of the Soviets with project Gemini, its second human space flight program designed to support the Apollo program. In 1965 they carried out the first rendezvous in orbit between two manned space crafts and a record endurance flight of fourteen days in space.

The immense heavy lifting rocket Saturn V was developed for the purpose of placing men on the Moon and bringing them back to Earth. The Saturn V program was directed by Arthur Rudolph, a rocket engineer who was one of the leaders of development of the V-2 rocket. The Saturn V rocket actually consisted of six spacecraft modules designed by five different companies. Three of the components were the rocket booster stages, the other three components were the command, service and lunar modules. All of this was made from several million individual components – that's over ten thousand times as many components as a typical mechanical Swiss watch, and it had to be just as reliable. For each Apollo mission, all of these modules were built new – they could not be flight tested, they just had to work.

Initially, von Braun's approach was to test every step and every change with launches – German thoroughness at its best. But the end of the decade goal would not be achieved at this pace, and eventually the entire Apollo-Saturn system was tested in flight without all the preliminaries. The most difficult part to construct was the lunar lander module. It had to carry the astronauts to the surface of the Moon from the orbiting command module and launch back into space to rejoin the command module. This task would be easier on the Moon due to its much lower gravity, but that also meant that it could not be fully tested on Earth. Designs began in 1962, and it was completed in 1968.

The Saturn V rocket is still the largest and most powerful rocket that has ever been developed. Ten metres in diameter, a hundred and ten metres tall, it could carry a payload of 140 tons to a low Earth orbit and a further 44 tons to the Moon. Over ninety percent of the weight of the Saturn V was fuel and fully fuelled it weighed almost three thousand tons. The first stage used kerosene, similar to jet fuel, the second and third stage boosters used liquid hydrogen, and they all used liquid oxygen as the oxidiser. The Saturn V was controlled by a computer developed in the early 1960s, which had the power of an early Apple II computer. But even though your smartphone is thousands of times more powerful than the Apollo control computer, the Apollo computer was designed to work in space and to have zero chance of an unexpected failure.

Space is not easy today, and it was far more difficult for the Soviet and American pioneers of the 1960's. During this decade eleven astronauts lost their lives, either in training exercises or during failures in space. The first Apollo mission – Apollo 1 in 1967 – ended in tragedy when a fire started in the astronauts' cabin during a test of the launch countdown. The pure oxygen atmosphere of the cabin caused the fire to quickly spread, asphyxiating the astronauts. The following Apollo missions began at number four to honour the crew.

Following Apollo 1, manned spaceflights were postponed for 21 months for thorough tests and a review of all designs. Later missions used a mixture of nitrogen and oxygen for the cabin atmosphere. Apollo 4, 5 and 6 were unmanned flights to test each phase of the Moon landing missions. Only the landing stage on the Moon could not be tested. Apollo 6 used the final version of the Saturn V rocket that would be used in all the next missions. In October 1968 Apollo 7 carried three astronauts into a low Earth orbit. Two months later Apollo 8 carried three astronauts to orbit the Moon and return to Earth. Apollo 9 tested other aspects of the final mission, whilst Apollo 10 was a full dress rehearsal, carrying out all procedures apart from landing astronauts on the Moon.

**Destination Moon**

On July 16th 1969, Apollo 11 launched from Kennedy Space Centre watched by a million people from nearby vantage points and over three thousand media reporters. The Saturn V first stage engines burned 13 tons of fuel per second but the start was slow – it took twelve seconds just for the giant rocket to move past the launch tower. The thrust generated by the engines was just enough to lift the rocket off the Earth's surface. But as fuel was burnt the weight decreased, and the constant thrust began to accelerate the rocket faster. That's the Tsiokolvsky rocket equation in action.

At an altitude of 67 kilometres the first stage rockets completed their burn, automatically detached and fell into the Atlantic Ocean 560 kilometres away from the launch site. The second stage took over, and the new engines fired for another six minutes to bring the rocket to an altitude of 180 kilometres and a speed of seven kilometres per second. That's about the speed needed to maintain a circular orbit around our planet for a rocket or satellite in a low Earth orbit. The second stage was ejected and landed over four thousand kilometres from the launch site.

After orbiting the Earth for three hours the spacecraft was pointed towards the Moon, and the third stage engines burned their fuel for six minutes. This gave the remaining spacecraft a speed of 11.2 kilometres per second – the escape speed from the Earth. After this the rocket was travelling quietly through space at a constant speed towards the Moon. At the top of the stage three rocket sits the lunar module landing craft and above that the command module with the astronauts.

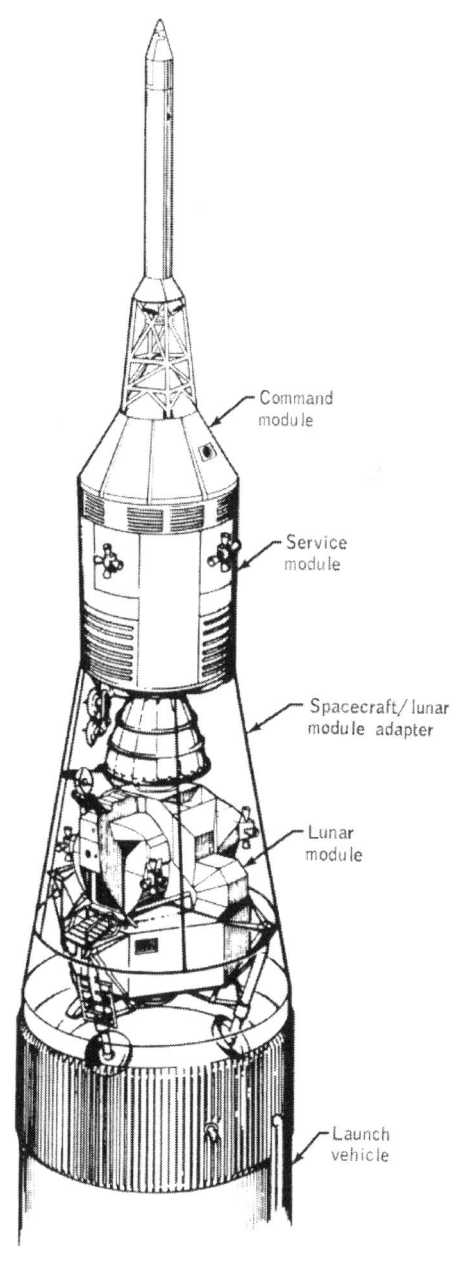

*Apollo launch configuration*

During its journey to the Moon the entire rocket was slowly rotated so that any one side would not spend too long in the heat of the Sun. As the rocket is drifting towards the Moon a complicated manoeuvre takes place. At the very top of the rocket is the lunar command module with the astronauts. Beneath that lies the lunar landing module. The command module must eventually fire its rockets to slow down and enter into an orbit around the Moon, but its engine is pointing towards the landing module beneath. On the way to the Moon it detaches, turns around 180 degrees and reattaches to the lunar module. This must have been a nervous time for the astronauts, but it was successful. Once completed, the command module and the landing module separated together from the stage three rocket. The stage three rocket was fired away, and the remaining components of the Saturn V rocket continued to the Moon.

Whereas the first two booster stages fell back to Earth, the stage three rockets were initially ejected into space and continued to orbit the Earth. In 2002 an amateur astronomer thought he had found a new asteroid orbiting the Earth. Astronomers looked in more detail and measured its reflected light spectrum. It turned out to have a signature of titanium dioxide – the paint NASA used to coat the Saturn V rockets.[11] In the later Apollo missions, the stage three rockets were deliberately aimed at the Moon so that their impact could be detected by the seismometers left during previous missions.

---

[11] Jorgensen, K. 2003, "Observations of J002E3: Possible Discovery of an Apollo Rocket Body", Bulletin of the American Astronomical Society, Vol. 35, p.981

On route to the Moon everything went to plan. The astronauts broadcast live colour images from the command module and received regular updates from Earth. They were already front page news all over the world, even in the Soviet Union. On July 19th, three days after launch, Apollo 11 passed behind the Moon and fired its engines to enter an orbit about the Moon.

Launched three days before Apollo 11, the Soviet Luna 15 robotic mission was designed to land on the Moon, retrieve samples of Moon rocks and return them to Earth. After Apollo 11 entered into a lunar orbit, the American mission control received a message from the Soviets congratulating them on reaching the Moon. They informed mission control that Luna 15 was currently orbiting the Moon and if it presented any problem they would move it. Performing a remote collection and return of lunar material was also an ambitious task. Unfortunately, Luna 15 crash landed about 800 kilometres away from the Apollo 11 landing site.

Luna 15 would have beaten the Americans to retrieving a piece of the Moon and bringing it back to Earth. The following year the Soviets were successful with their Luna 16 mission. An on board computer controlled its landing module, and then an automatic drill took seven minutes to penetrate 35 centimetres into the lunar regolith. Part of the landing craft then fired its ascent boosters and made the return journey to Earth carrying with it 100 grams of the Moon.

The two Apollo 11 astronauts Neil Armstrong and Buzz Aldrin entered the lunar landing module, which detached from the command module piloted by the third astronaut

Michael Collins. Its engines fired to bring it down to the surface of the Moon. The landing point was the Mare Tranquillitatis, the Sea of Tranquillity, chosen for its flat and smooth surface from the close up photographs taken by the Ranger and Surveyor missions. The landing module had enough fuel for six minutes of burn time.

During descent the landing module computer was sounding numerous different alarms. It was overloaded with information and not powerful enough to process all the data. There was an error in the check-list manual which left a radar switch in the wrong position. More alarms sounded, and Armstrong took over the controls. The landing site was covered with large boulders five metres high that were not visible on the earlier space photographs. Armstrong manoeuvred past to a smoother area. Despite many practice sessions on Earth in the flight simulator, Armstrong later said that it was far more difficult than any exercise. As they approached the surface, clouds of dust were blown up and blocked the view. With only 25 seconds of fuel remaining, touchdown was successful on July 20th 1969.

This phase of the Moon landing was completely untested and unknown. Aside from the concern that the lander might sink into metres of dust, there was another concern that the dust on the lunar surface might be pyrophoric and react with the oxygen inside the landing module, exploding when the door was opened. Fortunately, neither of these concerns were realised.

After skipping a scheduled sleep break and performing several hours of preparation, Neil Armstrong stepped down the ladders on to the surface of the Moon: "*I'm at the foot of the ladder. The LM [lunar module] foot pads are – only depressed about one to two inches, although the surface appears to be very, very fine-grained – almost like a powder – very fine. OK, I'm going to step off the LM now.*" Then, that famous phrase: "*That's one small step for [a] man. One giant leap for mankind.*"[12] Aldrin followed later and was a little more focussed on bodily activities: "*I decided to take that period of time to ah… to take care of a bodily function of slightly filling up the urine bag, so that I wouldn't be troubled with having to do that later on.*"[13] Much later we learned that when he jumped onto the surface of the Moon his urine collector broke, and one of his boots filled with pee!

Even before the launch he was down to Earth in his thoughts: "*I gave up smoking the pipe maybe three weeks before launch. Having a drink, three days before. I don't think anybody really slept too well the night before. We had what we call a faecal containment garment, which you wear underneath your underwear to contain any bowel movement throughout a prolonged flight, so, early in the morning, I greased the lower portion of my body with diaper-rash stuff.*"[14] After stepping onto the lunar surface he simply said, "*Beautiful, beautiful. Magnificent desolation.*"

Apollo 15 astronaut David Scott tells the difficulty in explaining to people what it is like to stand on the Moon: "*People always ask, what was it like? It's hard to describe because you can't take them there, and that's why I think, in the future, you*

---

[12] https://www.hq.nasa.gov/alsj/a11/a11.step.html
[13] Quoted from The Telegraph, 2008 article "What was it like to walk on the Moon?"
[14] Quotation from "In the Shadow of the Moon", 2007 documentary

*need an artist or poet or writer to go and express it for people in terms they can understand better."*[15]

All the Apollo explorations of the surface of the Moon took place during the long lunar days when the surface of the Moon is illuminated by the Sun. The Moon is smaller than the Earth, and the visibly curved horizon lies only about two and a half kilometres away, half as distant as the horizon on Earth. The landscape appears as shades of grey and it is difficult to perceive the distances and sizes of objects since there is nothing familiar with which to compare.

The lack of an atmosphere to scatter sunlight means that the sky is dark during both night and day. There are no colourful sunsets on the Moon, and because of the Moon's slow rotation, it takes an hour for the Sun to sink below the horizon. At the instant the last of the Sun moves below the horizon the landscape immediately becomes completely dark. But if the Earth were visible there would be a faint light from its reflection. The disk of the Earth is over thirteen times larger in the sky than the Moon appears from Earth. The continents and oceans would be clearly visible and the Earth would be seen to slowly pass through phases like the Moon.

The dazzling Sun prevented the astronauts from seeing the stars because their eyes were adapted to high light levels. That's the same reason why their cameras could not capture the stars – their exposure times were set for daylight conditions. If the astronauts had closed their eyes for several minutes to let them adapt to the dark, they would have been

---

[15] Quotation from "In the Shadow of the Moon", 2007 documentary.

able to see a spectacular starry sky and the Milky Way galaxy. Without an atmosphere, the stars would be brighter than seen from Earth, and they would not twinkle.

The Apollo 11 astronauts spent less than two hours exploring the surface of the Moon, walking up to a hundred metres from the landing craft, taking photographs and videos, performing planned experiments and collecting surface rocks and lunar dust to bring back to Earth. They collected 22 kilograms of material, including 50 rocks, samples of the fine-grained lunar 'soil' and two core tubes that included material taken from 13 centimetres below the Moon's surface.

Because of weight concerns, Apollo 11 took just two experiments to leave behind on the Moon. The more important one was the seismometer for measuring possible Moon quakes or asteroid impacts, so that the internal structure of the Moon could be inferred. The second instrument was the lunar retro reflector, a set of parabolic mirrors that could be used to reflect the beam of a powerful laser from Earth. The time for a pulse of laser light on its journey to the Moon and back would be later used to accurately measure the distance to the Moon.

After 21 hours on the surface of the Moon, the astronauts returned to the landing module, fired its ascent engines for seven minutes, which brought them back into a lunar orbit and docked with the orbiting command module where Michael Collins waited, orbiting the Moon alone.

Collins has been described as the world's loneliest man in the universe as he orbited the Moon 27 times alone in the command module. But he said: "*I discovered later that I was described as the loneliest man ever in the universe or something which really is a lot of baloney. I mean, I had Mission Control yakking in my ear half the time. Everything was going well with the command module, I had my happy little home, I had the bright lights on and everything was fine.*" He described the view from his orbiting module: "*When the Sun is shining on the surface at a very shallow angle, the craters cast long shadows and the Moon's surface seems very inhospitable. Forbidding, almost. I did not sense any great invitation on the part of the Moon for us to come into its domain. I sensed more that it was almost a hostile place, a scary place.*"[16]

**Homeward bound**

The lunar module was jettisoned and eventually crashed back onto the Moon. Even though in today's money this module cost over a billion dollars to construct, it was not designed to endure re-entry through Earth's atmosphere. It was deliberately allowed to crash onto the Moon so that the seismometers would detect the shock waves of the impact from an object of a known mass.

The command module burned its engines to bring it back towards Earth. On the journey home the astronauts took turns to broadcast their thoughts about the mission. Armstrong spoke of the Jules Verne novel about a trip to the Moon written just over one hundred years earlier, underscoring man's determination to venture out into the unknown and to discover its secrets. Just before re-entry into Earth's

---

[16] Quotation from The Telegraph, "What was it like to walk on the Moon" 2008.

atmosphere, the service module was jettisoned, and the small command module turned 180 degrees for its heat shield to face Earth.

Michael Collins described the re-entry through the atmosphere, when the command module roared back to Earth at up to 41,000 kilometres an hour: *"You are literally on fire. Your heat shield is on fire and its fragments are streaming out behind you. It's like being inside a gigantic light bulb."*[17] Atmospheric drag and eventually parachutes brought the final part of the Saturn V rocket to splashdown in the Pacific Ocean, eight days after leaving Earth.

The recovery team wore biohazard suits in case any pathogens had been brought back from the surface of the Moon, even though that was considered unlikely. The astronauts were taken to an isolation unit for two further weeks of quarantine. This was in accordance with the 'Extra-terrestrial exposure law', NASA regulations for guarding Earth against potential contamination by alien life-forms. This practice would continue for two more missions, Apollo 12 and Apollo 14, before the Moon was proven to be barren of life and the quarantine process dropped.

Apollo 11, 12, 14, 15, 16 and 17 landed a total of twelve astronauts on the Moon. Due to an accident Apollo 13 did not make it to the Moon. Two days into that mission an oxygen tank exploded and damaged the service module. In a dramatic series of events that has been the plot of several movies, the crew had to improvise and use the lunar module

---

[17] Quotation from The Telegraph, "What was it like to walk on the Moon" 2008.

to make the journey back to Earth. The Lunar Module was intended to support two people for a day and a half, not three people for four days. Their re-entry through the atmosphere was difficult, but the astronauts made it back to Earth safely.

In 1903 the Polish science fiction writer Jerzy Zulawski wrote *'On a silvery globe'* in which lunar travellers used a roving vehicle to explore the Moon. His foresight turned into reality during the last Apollo missions. The Apollo spacesuits and backpacks weighed 82 kilograms on Earth, but only 14 kilograms on the Moon. Still, it was hard work for the first astronauts to walk on the Moon in their clumsy space suits, and they could only explore a region the size of a football pitch. The idea of a lunar vehicle was proposed in 1969 and was developed very quickly.

The first use of the lunar vehicle during the Apollo 15 mission in 1971 allowed the astronauts to explore 27 kilometres of the Moon's surface. The rovers were very sophisticated vehicles, designed to work in the cold and heat of the Moon, climb slopes up to 45 degrees, carry two astronauts with steering on both sides of the vehicle and operate in a virtual vacuum. They were battery powered since conventional engines would not work without an oxygenated atmosphere.

The Apollo lunar rovers were not the first vehicles on the Moon. In 1970 the Soviets landed a solar power remote controlled rover in the Sea of Rains. Named Lunokhod 1 it was designed to work for one lunar day but kept functioning for eleven lunar days robotically exploring the surface of the Moon.

The last three Apollo missions all took lunar roving vehicles so they could explore more of the surface. These vehicles were left there and will no doubt become great attractions for future Moon tourists. The final mission was Apollo 17 in 1972, when Eugene Cernan was the last man to walk on the Moon. Three more Apollo missions had been planned but were cancelled for budget reasons. The job had been done.

Having been beaten to the Moon, the Soviets focussed on unmanned missions to Venus and Earth orbiting space stations. The first space station, Salyut 1, was launched in 1971 and hosted a total of six cosmonauts over its 175 days in orbit. Over the following two decades the Soviets placed six other space stations in orbit – several for military use. In a temporary thaw of the cold war, US president Richard Nixon and Soviet president Leonid Brezhnev agreed on a joint mission in which the last of the Apollo rockets was used to dock with the Soviet Soyuz craft in space in 1972.

The Apollo program had begun in 1961, and the first Moon landing took place just eight years later. It was an incredible achievement in such a short timescale. Every aspect of the journey, landing and return had to be designed and developed. It shows what we are capable of given motivation and resources.

From the mid-1960s to the mid-1970s there were 45 successful Moon missions. From 29 Soviet missions 21 were successful, and from 29 US missions 24 were successful. The pace of exploration was intense and is far above today's space program – there were 10 missions alone in 1971. But after

Luna 24 in 1976 they suddenly stopped. The two-nation space race was seldom mentioned again, except by those who doubted that the Soviets had ever intended to send men to the Moon. The United States focused on Mars and the outer solar system and on the Skylab and Space Shuttle programs. The Soviet Union dissolved in 1991, and the remains of its space program mainly passed to Russia. Since then the United States and Russia have worked together in space with the Shuttle–Mir Program and again with the International Space Station.

The long lasting legacy of the Apollo missions was the human achievement and allowing people on Earth to see their planet in a new way. All of the astronauts who visited the Moon remarked on the spiritual feeling of seeing the Earth in its entirety, a small home in a vast universe. The scientific legacy was seen as second place to beating the Soviets. But it was a very important legacy. Only four nations did not cover the Moon landings: China, Albania, North Korea, and North Vietnam. Perhaps the only person on the planet who was not impressed was Pablo Picasso, who in an interview in The New York Times said, *"It means nothing to me. I have no opinion about it, and I don't care."*

## 3. The Apollo Legacy

The Apollo program is regarded as one of the most important events of the 20th century. Incredibly dangerous and pioneering, it was a symbol of human achievement that pushed the existing technology to its limit. America achieved its goal of demonstrating its dominance in space – a legacy that continues until this day. Soviet Nobel laureate Andrei Sakharov and his colleagues issued an open letter to the Soviet government in 1970, calling for democratisation of the USSR and citing the American Moon landing as evidence of the superiority of democracy.[18] The technological legacy was also important. There were numerous spin-offs from the program, from the Earth monitoring program Landsat to freeze dried food, flame resistant materials and much more.

The scientific legacy of the Apollo program cannot be understated. Only by viewing the Moon up close, sampling its surface and returning with pieces of the Moon to Earth could we learn about the history of the Moon, our Earth and the solar system. The initial ideas of Harold Urey, the key science advisor to NASA for the Apollo program, were realised. Even today, fifty years after Apollo 11, scientists are still uncovering the secrets of our Moon thanks to the experimental data that was gathered.

In the three and a half years between the first and last footprints on the Moon, twelve astronauts spent a total time of twelve and a half days on the lunar surface. Lunar samples

---

[18] "The need for democratization", 1970, A.D. Sakharov, V.F. Turchin and R.A. Medvedev, The Saturday Review, June 6, page 26-27

were collected from six landing sites, measurements were made, and instruments were deployed that revolutionised lunar and planetary science and continue to have a major scientific impact today.

The six Apollo landing sites were all near the central region of the Moon that faces Earth. The polar regions of the Moon were unexplored because of the added complexity of returning to the command module, which orbited the lunar equator. From the lunar equator it is far easier for the command module to take what is known as a 'free return trajectory' to Earth. This minimises the fuel needed for the flight back to Earth by using the Moon's motion and gravity to provide energy. No astronaut has walked on the far side of the Moon because radio communication with Earth would have been impossible without a satellite relay system.

Each of the Apollo missions took a scientific package to be left on the Moon. They left seismometers to measure the internal structure of the Moon. Experiments were carried out to determine the particle composition of space near the Moon and the interaction of the Moon's magnetic field with that of Earth. There were experiments to measure any residual atmosphere on the Moon, as well as to capture and determine the composition of the solar wind. Laser ranging reflective mirrors were left behind. An ultraviolet camera was positioned to return photography of the Earth and stars in spectral bands that are blocked by Earth's atmosphere. Experiments were made to measure the flow of heat from the inside of the Moon. Biomedical experiments included the Biostack experiment to study the effects of cosmic rays on living organisms, the BIOCORE experiment that measured the effects of high energy particles on the brain and MEED, which examined the effects of ultraviolet radiation and the vacuum of space on microorganisms. Small satellites were placed in orbit around the Moon to investigate its topology, mass and variations in its gravitational field and to measure the infra-red radiation from the Moon to determine how quickly its surface cooled at night.

The Apollo landing sites were near to or within the largest lunar craters, and rock samples were brought back from each location. Three of the Soviet Luna missions also achieved unmanned landings on the Moon, and these robotic missions brought back a small sample of the Moon's surface to Earth. Some missions extracted core samples of the Moon from one metre below the surface. On foot and in roving vehicles, 95 kilometres of the lunar surface was explored, 382 kilograms of lunar regolith and rocks were recovered, and over two tons of scientific equipment were left behind. Monitoring experiments were also conducted from the orbiting command/service modules. So, what did we learn from all of this?

There were many highlights of scientific discovery that followed. The primary findings included determining the age of the Moon and its surface features, measuring its onion-shell like internal structure and finding a small iron core, determining the history of cratering over the past four billion years, measuring that the Moon moves away from Earth a few centimetres each year, the discovery that the Moon is made of material identical to the Earth, and that the far side of the Moon is strangely different from the near side. The Apollo missions also confirmed the volcanic origin of the vast lunar mare and the impact origin of the lunar craters, as well as the absence of life on the Moon.

**Inside the Moon**

Neal Stephenson's 2015 book '*Seven Eves*' is an epic hard science fiction story of the future of humanity. It begins "*The moon blew up without warning and for no apparent reason.*" However, it is difficult to imagine a scenario in which our Moon could disintegrate. The Moon is a giant sphere of rock and metal with a mass of 73 quintillion tons and a diameter of 3,474 kilometres, just over a quarter of that of the Earth. Its volume is therefore about two percent that of the Earth, but its mass is just 1.2 percent of the Earth because the average density of the Moon is less than the Earth's. Its average distance from us is just over 30 Earth diameters away. But because the Moon makes an elliptical orbit about the Earth, at its closest approach it is 28.5 Earth diameters away and at its most distant it is 31.8 Earth diameters away. To gain an intuitive idea of the size of the Earth and Moon, if the Earth were a basketball the Moon would be the size of a tennis ball. To illustrate the distance to the Moon it may surprise you to learn that the tennis ball should be placed about seven metres away from the basketball.

Like our Earth, the Moon has a crust, mantle and core, but the size and compositions of these regions are rather different. This so called 'differentiated structure' of a core, mantle and crust could only arise if the Moon was once completely molten. The internal structure of the Earth can be reconstructed by modelling its density and by studying the sound waves made by earthquakes as they propagate through its interior. Some of these acoustic waves reflect off the iron core of the Earth.

The outer regions of the Moon are thought to be made of cold solid rock – that's why the Moon does not have plate tectonics or recent active volcanism. Neither does it have 'Moonquakes' due to collisions of surface plates. But as the Moon orbits the Earth, it is squeezed by Earth's gravity, and that causes its internal structure to shift and crack. This causes Moonquakes deep within its interior, and their vibrations can be analysed using the seismic detectors left on the Moon at different locations by the Apollo astronauts. These Moonquakes occur with striking regularity – monthly trembles that are related to tidal stresses caused by the elliptical orbit of the Moon about the Earth.[19]

From studying the Moon rocks we have learned that after its formation the Moon's mantle was molten hot – a giant sea of magma. A geological process called fractional crystallisation would have occurred as the Moon radiated its heat into space and cooled. Solid crystals of minerals like olivine and anorthosite formed within the cooling magma, and the heavier rocky and iron rich crystals would sink towards the core whilst the lighter crystals rose to form the crust.

The radioactive decay of elements in the Earth and Moon provide an important source of heat, which kept their inner regions hot over billions of years. But there was another

---

[19] Kawamura, T. et al. 2017, "Evaluation of deep moonquake source parameters: Implication for fault characteristics and thermal state", Journal of Geophysical Research, 122, 7, 1487.

source of heat for our Moon. Early in the Moon's history it was much closer to Earth, just a few Earth diameters away in fact. This was figured out by George Darwin, the son of Charles Darwin, and we will learn more about his work in later chapters. The tidal stresses on the Moon were enormous – so large that the Moon would have been continuously deformed and squeezed into an ellipsoidal egg shape. Because the Moon's orbit is not an exact circle, this would have caused material inside the Moon to be continually squeezed and stretched, heating its insides.

Although scientists have long debated what lies at the centre of the Moon, in 2011 a careful reanalysis of the old Apollo seismic data revealed that the Moon has a relatively small iron core with a radius of about 200 kilometres.[20] This iron sphere is just one percent of the mass of the Moon, compared to Earth's metal core, which is over a third of the mass of our planet. That's quite a difference and explains why the Moon is less dense than the Earth. But more importantly, it is a significant clue about the formation history of our Moon – that there was far less iron in the material from which it originated.

The Moon's solid iron core is surrounded by a thin hot liquid metal boundary layer about a hundred kilometres wide, and beyond that lies the large surrounding mantle. The mantle is thought to be partially melted in the lower regions next to the core, but most of the mantle is now cold solid rock. That's because the Moon is smaller than the Earth, and small bodies cool down more quickly than larger bodies. This

---

[20] Weber, R.C. et al. 2011, "Seismic detection of the lunar core", Science, 331, 309.

happens because the amount of heat stored in the interior is proportional to the volume, but the rate at which heat is radiated into space is proportional to the surface area. This means that larger objects store more heat for a longer period of time.

The Moon cooled quickly, the crust and outer regions solidifying in a hundred million years or so. The lunar highlands are the oldest material on the Moon – they solidified from the magma ocean 4.4 billion years ago. The anorthosite crust of the Moon is estimated to be about fifty kilometres thick, comparable to Earth's solid crust but much thicker relative to its size. Parts of the surface are coated with basalt rock, where the molten insides of the Moon once spilled onto the surface and solidified, forming the darker grey lunar Mare.

**A dangerous place to stand**

The most obvious features on the Moon are the numerous craters that cover its entire surface. Over time, countless smaller asteroids, meteoroids and space dust have collided with the Moon, covering its entire surface with craters. The craters overlap, and there are smaller craters within the larger craters.

In the 19th century, the astronomer Julius Schmidt estimated that there are 30,000 craters on the Moon. This number was limited by the resolving power of telescopes at the time. There are about 5000 craters on the Moon larger than 20 kilometres across and over half a million craters larger than one kilometre. And the numbers keep increasing for smaller sizes – the entire surface of the Moon has been impacted by some space rock at some point in time, and most places have been hit multiple times.

Since 1919, the International Astronomical Union has been the arbiter of planetary and satellite nomenclature. Over nine thousand features on the Moon have been named, including the lunar seas (mare), craters, mountains (montes), ridges (rilles) and valleys (valles). But there are far more that are nameless. If you find a nice feature on the Moon that doesn't have a name and wish to honour someone you can submit a proposal to the International Astronomical Union. As of 2018, 1619 craters have been named.[21] But there are a vast number that have not been named – there are enough craters larger than 10 metres across to name everyone on Earth!

Our Earth has very few impact craters on its surface. The reason that the Moon looks so different from the Earth is the lack of an atmosphere and the lack of plate tectonics. These processes on Earth lead to a surface that is relatively new – there is no weather on the Moon, no wind, rain or rivers, nor are there movements of landmasses that lead to mountains and volcanoes. The Moon's surface is a record of the last four billion years of history of our solar system.

Once it was realised that the lunar craters were due to asteroid and meteoroid impacts, the impact frequency could be measured. That's because if one crater lies within another, or overlaps with its ejecta with another, it must be the younger crater. But to determine the impact rate over time it was necessary to measure real ages for some of the craters. This was one of the major scientific accomplishments of the

---

[21] https://planetarynames.wr.usgs.gov/Page/MOON/target

Apollo program – once the ages of several craters were determined by dating Moon rocks from the craters, the counts of craters within craters could be used to determine the impact rate on the Moon over the past four billion years. It's an important result because the cratering timescale of the Moon is used to estimate the surface ages of different features on other solar system planets and asteroids.

The claim that our Moon preserves a record of our solar system quite unlike our Earth is clear when we realise that the Tycho crater, with its bright streaks emanating outwards, formed from an impact 100 million years ago – an impact that would have been witnessed by the dinosaurs on our planet. And the Copernicus crater that formed 800 million years ago occurred before the Cambrian explosion – at that time there was no creature on Earth with eyes to witness that cosmic collision.

The Moon is still being bombarded today, although at a much lower rate than in the past. For the last couple of decades, several astronomical observatories have had video cameras trained on the Moon. They are searching for the bright flashes of light that occur when small asteroids impact its surface. Dual telescopes are used, so an impact event must be seen at the same time in both cameras to register an event. In this way, hundreds of confirmed new lunar impacts have been detected each year.

Around 100 tons of space dust and sand sized particles hit the Earth every day, but anything smaller than a few metres across burns up harmlessly in our atmosphere. On the Moon there is no protective atmosphere, and space rocks hit the surface travelling at speeds between 20 and 70 kilometres per second. Even a small space rock weighing five kilograms can create a crater on the Moon about ten metres across, scattering over 70 tons of lunar regolith in all directions.

The cameras trained on the Moon can detect objects as small as a few grams hitting the surface. That may sound difficult, but one gram travelling at 70 kilometres per second has an energy equal to one kilogram of TNT explosive. To relate the brightness of the flashes to the size and speed of the impactor, experiments on Earth have been carried out. Artificial Moon rocks are created and a hypervelocity gun shoots projectiles at the rocks at speeds up to six kilometres per second, the highest achievable by a gun. The resulting flashes from projectiles of different sizes are then compared to the observed flashes on the Moon.

The brightest impact flash was observed in 2013 within the Mare Imbrium crater. The coordinates were passed onto the Lunar Reconnaissance Orbiter, a NASA satellite that has been orbiting and mapping the Moon since 2009. The satellite had previously taken high resolution images of this region, and a new set of images revealed the brand-new impact crater, twenty metres across with debris scattered in all directions. Over 250 secondary impacts were detected, some 30 kilometres away from the primary impact crater.[22] This illustrates one of the potential dangers of an inhabited and exposed lunar base.

---

[22] Robinson, M.S. 2015, "New crater on the Moon and a swarm of secondaries", Icarus, 252, 229.

A 2016 Lunar Reconnaissance Orbiter study compared images from several years apart and found hundreds of new craters. Craters larger than 10 metres across are being formed at a rate of an average of 16 per year. Together with the ejecta that such impacts spread across the lunar surface, it was recently calculated that the top two centimetres of lunar regolith is churned up every 81,000 years.[23] This is a hundred times higher than older estimates, so the Apollo astronauts' footsteps on the Moon will most likely not survive for millions of years as often quoted.

If the ejected material from crater formation reaches speeds of 2.4 kilometres per second, it escapes the Moon's gravity and begins orbiting the Earth or Sun. Over time, some of these ejected Moon rocks land on Earth as meteorites. A good place to find meteorites is Antarctica, where they lay exposed on ancient ice sheets. Around 37,000 meteorites have been recovered from Antarctica, of which 35 have been confirmed to originate from the Moon![24]

About one in a thousand meteorites originate from the Moon, and a similar number comes from Mars. The rest come from asteroids and meteoroids that have been orbiting the Sun for billions of years. Unfortunately, no meteorites are known to have come from Venus because a sample of Venus would solve a debate about the Moon's formation that I will explain in the next chapter. The gravity of Venus is similar to Earth and it has a thick atmosphere, a hundred times as dense as Earth's. This gaseous shield prevents anything but the

---

[23] Speyerer, E.J. 2016, "Quantifying crater production and regolith overturn on the Moon with temporal imaging", Nature, 538, 215.

[24] http://meteorites.wustl.edu/lunar/moon_meteorites.htm

largest asteroids from producing craters, and it also prevents any ejected material from escaping its gravitational pull.

Over one hundred distinct meteorites are known to have originated from the Moon – that's over 200 kilograms of material. Scientifically, these meteorites are important because they provide additional material from different regions of the Moon to analyse. However, they have also suffered from spending decades in space being bombarded by cosmic rays and the final destructive flight through Earth's atmosphere.

How do we know whether a meteorite is a random space rock or comes from the Moon? All meteorites contain certain isotopes that can only be produced by reactions with penetrating cosmic rays while outside the Earth's atmosphere. The presence of so called 'cosmogenic nuclides' is the ultimate test of whether or not a rock is a meteorite. And because some of these cosmic-ray induced isotopes are unstable, it is possible to tell how long ago since a meteorite fell to Earth and how long it has been orbiting in space. The shortest journey time for a lunar meteorite was one hundred years whereas another rock took over ten million years to arrive.

### A dusty surface

The patches on the near side of the Moon that you can see with your eyes are due to the darker basalt rock of the lunar maria[25]. They formed from molten mantle material which flowed onto the lunar surface billions of years ago, covering

---

[25] Maria is the plural term for mare.

pre-existing giant impact craters. From the Moon rocks brought back from these regions, it was discovered that all of the largest craters on the near side of the Moon formed in a narrow interval of time around four billion years ago.

This period in the Moon's history is called 'the late heavy bombardment'. It may have been the tail end of a previously higher rate of impacts. Or it may have been a one-off cataclysmic event, perhaps caused by an influx of asteroids and comets from the outer solar system, sent inwards by the orbital motions of the giant planets Jupiter and Saturn. To understand the early history of our Moon in more detail we would need to analyse Moon rocks from a diverse range of locations across its surface.

The lunar Mare are surrounded by the lighter coloured rocks of the ancient crustal highlands. The tallest mountain on the Moon is Mons Huygens which is 5.5 kilometres high and lies at the edge of the giant Mare Imbrium. But the lunar mountains are not like the mountains on Earth that are made by colliding surface plates. The lunar mountains are the walls of craters, or material that was pushed aside during giant asteroid impacts.

The highest point on the Moon, the Selenean summit, lies on the far side of the Moon. Rising 10,786 metres above the mean surface height of the Moon, it is nearly 2000 metres higher than Mount Everest! It was discovered in 2010 by the Lunar Reconnaissance Orbiter, which accurately mapped the surface topology of the Moon using a laser altimeter. This lunar high point probably formed from molten ejecta from the giant impact that led to the South Pole Aitken crater. The first future Moon visitor to climb the Selenean summit would find it a relatively easy task since the entire hike would be up a very modest three-degree incline.

Shortly after the formation of the Earth and Moon, the solar system was still an extremely hostile place, filled with far more asteroids than today. The impact rate onto the newly formed Earth and Moon was intense. Over time, the number of asteroids decreased as they were accreted by the planets, fell into the Sun or were scattered into the outer solar system by gravitational interactions with Jupiter and Saturn.

In the billion years following the late heavy bombardment, volcanism and pyroclastic eruptions launched material through the near-side lunar crust filling many of the enormous craters that had formed from giant impacts. From crater counting it seems that some eruptions of magma material onto the surface may have occurred as recently as a billion years ago.

There are also some hints of recent activity on the surface of the Moon. In 2014 NASA announced evidence for recent lunar volcanism at 70 irregular Mare patches identified by the Lunar Reconnaissance Orbiter, some less than 50 million years old.[26] This raises the possibility of a much warmer lunar mantle than previously believed, at least on the near side where the deep crust is substantially warmer – perhaps because of a mysteriously high concentration of radioactive elements.

---

[26] Braden, S.E. 2014, "Evidence for basaltic volcanism on the Moon within the past 100 million years", Nature Geoscience, 7, 787.

It is not known how these lava eruptions occurred or why most of the hot lava emerged from the near side of the Moon and not the far side. The largest impact crater, and the second largest known in the entire solar system, is the South Pole-Aitken basin. This enormous crater on the far side of the Moon is two and a half thousand kilometres in diameter and thirteen kilometres deep. Yet it has not been filled in with volcanic magma.

The lunar maria regions contain long meandering narrow features called rilles. These often radiate from old volcanic vents and are thought to be ancient lava channels. There are also lava tubes as revealed by a number of pits and cavities by the high-resolution images. Some have skylight-like openings to space. These have been proposed as ideal locations for future underground lunar cities, shielded from micrometeorite impacts and the high energy particles from the Sun.

The impacts of asteroids have pulverised the surface of the Moon, producing a fine-grained surface layer called the lunar regolith. This layer can be tens of metres deep and very different from the surface layer of sand or soil on the Earth.

Despite concerns that the Apollo landing craft would sink into the surface layer of dust on the Moon, the lunar regolith actually provided good support. This meant that the final step onto the surface of the Moon was actually a large leap of over one metre since the lunar lander did not sink into the surface as much as expected! Whereas the dust on the surface of the Moon has resulted from the disintegration of anorthosite and basalt rocks by impacts and high energy particles from space, the Earth's surface layer has formed from tectonic activity, erosion and biological processes.

The lunar dust is one of the primary dangers for future lunar missions or a Moon base. The lack of an atmosphere and magnetic field allows the Moon's surface to be perpetually bombarded by the solar wind – very fast-moving nuclei and charged particles from the Sun. This exposure causes the dust to become electrostatically charged, and this charge can be so strong that the dust particles actually levitate above the lunar surface. The electrostatically charged particles clung to the astronaut's space suits and were subsequently carried back into their living quarters on board the lunar lander. Most of the lunar astronauts commented about the pungent smell of the lunar dust, which resembled that of gunpowder or wet ashes.

The day after Apollo 17 astronaut Harrison Schmitt walked on the Moon he accidentally breathed in some of the Moon dust that had entered the Moon lander. Schmitt suffered what he described as a lunar hay fever – watering eyes, a sore throat and constant sneezing. The lunar dust is not weathered like dust on Earth, and even the smallest particles are jagged and sharp, rather like fragments of shattered glass. This can cause significant damage to equipment and our bodies.

In a recent study it was found that tiny fragments of lunar dust might be transported into the brain via the olfactory bulb, where it could lead to neurological disorders.[27] In 2018

---

[27] Krisanova, N. et al. 2013, "Neurotoxic Potential of Lunar and Martian Dust", Astrobiology, 13, 679.

scientists studied the toxicity effect of lunar dust to the DNA of mammalian lung cells.[28] These alien dust particles were found to be capable of causing significant DNA damage and even the death of cells upon exposure, potentially leading to cancer.

## The far side

Thousands of years ago the Moon was considered by some to be a portal into a ring of fire. Others thought it was a perfect crystalline sphere. Even after the Moon was known to be a spherical object and its near side was mapped by astronomers, the far side of the Moon remained a mystery.

This all changed when the Soviet space probe Luna 3 orbited the Moon in 1959, took photographs and beamed the images back to Earth. It was another eight years before the astronauts of the Apollo 8 mission flew around the Moon and the far side was seen with human eyes. Moon globes before this period were blank on their back side and are collectors' items today. Photographing the Moon with an unmanned probe was a remarkable accomplishment at the time. Luna 3 was a cylindrical probe just over one metre in length, laminated with solar panels and bristling with antennae. It was controlled via radio from Earth, and its orientation could be changed using small jets of gas.

There were no digital cameras, and photographs were taken with 35mm film. Since the Luna 3 craft was not designed to return to Earth, the film had to be automatically

---

[28] Caston, R. et al. 2018, "Assessing Toxicity and Nuclear and Mitochondrial DNA Damage Caused by Exposure of Mammalian Cells to Lunar Regolith Simulants", Geohealth, 2, 4, 139.

chemically developed into a negative in a miniature dark room. The negatives were then moved into a chamber, where a cathode ray beam was projected through the film, and the dark and bright regions were recorded with a photoelectric multiplier. The resulting data was converted into an electrical signal that was beamed back to Earth using a frequency modulated analogue signal – essentially a fax sent via radio. Each image took around 30 minutes to be transmitted, and around a dozen wide angle and close up photos of the far side of the Moon were received before contact was lost with Luna 3.

The origin of the film used by the Soviets is itself an interesting story. In 1956, before the advent of spy satellites, the USA began secretly taking images of the Soviet Union using cameras carried on high altitude helium filled balloons. These balloons reached altitudes of up to 30 kilometres, where the cold temperatures and radiation from cosmic rays would have destroyed normal photographic film. The United States developed special space resistant film just for this purpose. The Soviets had not yet mastered the ability to produce film that could withstand being in space, but they had shot down and captured some of the American spy balloons that contained undeveloped film. And that is what they used in the Luna 3 camera.

The first photographs of the far side of the Moon were somewhat blurry, but they caused great excitement when revealed to the public. They showed a heavily cratered surface that was strikingly different from the near side of the Moon. Whereas about 35 percent of the near side surface of the Moon is covered with those darker grey patches, the volcanic maria, on its far side only 5 percent of the surface has been volcanically resurfaced.

The reason for this is not known. However, a big clue came from later lunar missions that mapped the gravitational field of the Moon. The crust of the Moon is about 15 kilometres thicker on the far side than on the near side. This must have made it easier for the volcanic magma from the mantle to break through to the surface. Why the crust is thinner on the near side is still debated. One theory suggests that the crust is thicker on the far side because a second ancient moon of Earth slowly collided with the Moon, sinking into the far side, and thickening its crust. Another idea is that because of the tidal interaction with Earth, there is a build-up of radioactive elements under the crust on the near side which provides a source of heat. Another theory maintains that the near side of the Moon basked in the hot glow of the early Earth, preventing its crust from thickening – this is also plausible since after its formation the Moon was much closer to the Earth.

*The near and far side of the Moon*

By now the entire Moon has been mapped to a resolution of two metres or less. NASAs Lunar Reconnaissance Orbiter has taken over 10,000 high resolution images, some of which can resolve surface features as small as fifty centimetres. These have been stitched together to create one of the world's largest images. This map of the Moon is online, consists of 867 billion pixels and takes up over three terabytes of storage space.[29] The data has enough resolution for you to find the Apollo landing sites and zoom in to see the base of the landing modules and the kilometres of tracks taken by the lunar roving vehicles. You can even see the tracks of the astronauts themselves, intact fifty years since those footsteps were taken.

**How old is the Moon?**

The oldest dated rocks in the solar system are meteorites, which – apart from those that have been shattered off the Moon and Mars – are all around four and a half billion years old. Their ages can be accurately determined to within a million years using radiometric dating. Certain isotopes of uranium decay into lead in a timescale of billions of years, and the age of a rock can be found by measuring the amount of uranium and lead. The age of the oldest meteorites is used to mark the birth of our solar system, and the oldest fragments that have landed on Earth as meteorites are 4.57 billion years old.

---

[29] http://wms.lroc.asu.edu/lroc , https://lunar.gsfc.nasa.gov/

On Earth, the oldest surviving rocks are just four billion years old, but there are numerous ancient zircon crystals that are up to 4.4 billion years old. Crystals form by incorporating atoms from the rocky matrix within which they are embedded. The mere existence of these ancient zircon crystals tells us that the Earth must have had a solid rocky crust 4.4 billion years ago. However, these first rocks may have been pulverised by subsequent impact events, leaving behind the tiny tough crystals as a record of their existence.

What about our Moon? The oldest surface rocks were expected to be the anorthosite on the lunar surface, which rapidly crystallised from the hot magma ocean soon after the Moon formed. However, samples of this rock returned to Earth revealed a younger than expected age of 4.3 billion years. Because many radiometric dating techniques measure the age of material from the time it solidified, this younger age is likely due to subsequent impact events that re-melted the surface of the Moon.

But, the anorthosite on the Moon also contains crystals of zircon. These small millimetre sized crystals are extremely tough and would have survived any impact events. In 2017, zircon fragments from the Apollo 17 Moon rocks were dated by measuring the abundances of uranium and lead. They found that the zircon crystals, and therefore the rocks in which they must have formed, are 4.51 billion years old.[30] This implies that our Moon formed just 60 million years after the

---

[30] Barboni, M. 2017, "Early formation of the Moon 4.51 billion years ago", Science Advances, vol 3, 1.

birth of our solar system – another important result that provides a clue as to the origin of our Moon.

As a check on these estimates for the birth of our solar system we can also calculate the age of our star. That might sound difficult since we do not have a piece of the Sun to measure. So how can astronomers determine the age of our star which has never been sampled? It turns out that stars are reasonably simple things – giant spheres of gas that undergo nuclear fusion at their centres. This transforms lighter elements into heavier elements and releases energy as a consequence. As a star ages, it changes its size, composition, temperature and brightness in a predictable way. This means that measuring a few simple properties of a star allows us to estimate the time at which it began to shine, and that time for our Sun was 4.6 billion years ago.

**The Earth and Moon are siblings**

Perhaps the most surprising discovery from the analysis of Moon rocks was that they are all identical to rocks on Earth. How do we know that, since rocks on Earth come in a vast range of types, ages and compositions? Not all atoms of an element are the same. Atoms have the same number of protons as electrons, but the number of neutrons that bind to the protons can vary. These variations of an element are called isotopes. The trick is to measure the abundance of different isotopes of certain elements. For example, oxygen is a common element within rocks, and oxygen can occur in three different stable isotopes, with 16, 17 and 18 neutrons.

During the early formation of our solar system, these isotopes ended up having different abundances in different regions of the proto-solar system. Within asteroids, comets or planets that formed at large distances from the Sun, the relative amount of these isotopes is different from those objects that formed closer to the Sun. This means that any rock on Earth will have the same relative abundances of these isotopes, but the abundances of these isotopes are very different to rocks on Mars, or within meteorites and comets.

As another example, a hydrogen atom usually consists of one proton and one electron. Its formal name is protium. A tiny fraction of hydrogen atoms contain a neutron bound to their proton. This heavier isotope of hydrogen is called deuterium. The deuterium fraction was recently found to be much higher in comets than on Earth, implying that Earth's water came from a different source. It is now thought that water on Earth originated from asteroids that formed in between the orbit of Mars and Jupiter.[31] The deuterium fraction in water within these so called 'chondritic' asteroids is a close match to the water on Earth and to the water on the Moon – more about that shortly.

The relative abundances of the three oxygen isotopes have been measured from the rock samples returned from all the different Apollo landing sites. The remarkable finding was that to an accuracy of better than 99.998 percent, the abundance of these isotopes was the same in lunar rocks and Earth rocks. This has the fascinating implication that the Moon must have formed from material that was once inside the Earth.

---

[31] Altwegg, K. 2017, "67P/Churyumov-Gerasimenko, a Jupiter family comet with a high D/H ratio", Science, 347, 6220.

The same results have been found with other isotope indicators, such as titanium and potassium, which gives even stronger support to the idea that the Moon originated from our Earth.[32] However, recent studies from 2018 using more sensitive equipment have found that there are very small differences in the abundance of these isotopes in Moon and Earth rocks, but only at a level of about three parts in a hundred thousand.[33] That's an important result because this tiny difference could arise from asteroids and comets that impacted the Earth after the Moon had formed – slightly altering the abundance of elemental isotopes via the material that arrived from the outer solar system.

This is just one example of a great deal of ongoing research taking place on the original Apollo Moon rocks. It might be time for NASA to release some of their sealed samples for further analyses – there are three containers of Moon rocks from Apollo 15, 16 and 17 that remain unopened. These samples were left sealed to avoid contamination with our atmosphere and to be available at a time when analysis techniques had improved.

To prevent the stored Moon rocks from contamination by Earth's atmosphere or bacteria, each sample is stored inside a stainless-steel container filled with inert nitrogen gas. The containers are kept within a vault isolated by an air lock and

---

[32] Pahlevan, K. 2014, "Isotopes as tracers of the sources of the lunar material and processes of lunar origin", Philosophical Transactions of the Royal Society A, 372, 2024.
[33] Greenwood R.C., et al. 2018, "Oxygen isotopic evidence for accretion of Earth's water before a high-energy Moon-forming giant impact", Science Advances, vol 4, 3.

a 14 ton door with two separate combination locks. The vault itself is monitored by guards, motion sensors and video cameras. And part of the collection is stored with the same care at a separate location in case of a natural disaster.

Why the security? Moon rocks are far rarer than diamonds, and even the Apollo astronauts were not allowed to keep any samples. Over 100 tiny pieces of Moon rock were given as goodwill gifts to nations of the world by President Richard Nixon. Many of these have since gone missing. In the 1990's NASA worked with an undercover agent to recover some of the lost Moon rocks. An advert was placed in a major newspaper under the headline 'Moon rocks wanted'. The owner of the 'lost' Honduras one-gram goodwill Moon rock took the bait and offered it to the agent for a price of five million dollars. It was subsequently returned to NASA. And if you like treasure hunting, Ireland's gift of an Apollo 11 Moon rock lies underneath a rubbish dump – thrown out by accident after a fire in the observatory where it was being kept.

Neil Armstrong's lunar sample return bag containing traces of Moon dust was sold for 1.8 million dollars at auction in 2017. NASA tried to prevent the sale, but the courts deemed that it had been purchased legally. The true price of Apollo Moon rocks is not known because there are no samples that can be legally sold. In 1993 an anonymous collector paid 442,000 dollars for 0.2 grams of the Soviet lunar dust returned by one of their Luna 16 robotic mission in 1970. This very same sample was sold at auction in 2018 for 855'000 dollars!

The above prices are all far above the collection and return costs. The Apollo program cost around 100 billion dollars in today's money, so the price of collecting 382 kilograms of the Moon was about 262,000 dollars per gram. Perhaps this is one way to fund a future mission to the Moon! However, you can purchase Moon rocks for less money. In 2018, a large 5.5 kilogram lunar meteorite discovered in the North African desert fetched 612,500 dollars at auction.

Some lunar researchers argue that now would be a good time to open one of the sealed containers of Apollo rocks, not for their resale value, but because their ability to measure trace amounts of water and gases has greatly improved since the 1970's, and another recent analysis of Moon rocks has revealed the presence of water. NASA has recently announced that they will indeed make some of their sealed collection available to scientists for analysis.

**Water on the Moon!**

Although many early astronomers thought that the dark lunar maria were giant lakes, once the Moon was discovered to have no atmosphere or weather, it was expected to be a dry and lifeless world. This was confirmed by the first Soviet and American missions to the Moon in the 1960's. Certainly, liquid water cannot exist on the surface of the Moon. Exposed to sunlight and in the vacuum of space the water would be split apart into hydrogen and oxygen. The hydrogen would escape into space and the oxygen would bond to minerals on the surface – rather similar to what has happened on Mars, which was once a world with abundant flowing water.

Although trace amounts of water were found in the Apollo rocks when they were first analysed back on Earth, it was thought that this was due to contamination. But more recent measurements have revealed that the Moon is not completely devoid of water. In 2008, another careful reanalysis of the Apollo rocks found traces of water trapped in volcanic glass beads that could not have been contaminated.[34] In 2013 another study of crustal rocks found traces of hydroxyl, a molecule of hydrogen and oxygen, which suggested that the ancient rocks from the lunar highlands contain water at a level of six parts per million. From these measurements it is estimated that the lunar surface could have once contained over one percent water by weight.

There is also some recent evidence that large amounts of water ice exist deep underneath the lunar surface. This comes from the analyses of meteorites on Earth originating from material that was blasted off the Moon during the formation of lunar craters. One of the lunar meteorites contained the mineral moganite, which can only form in the presence of water under high pressure conditions.

Already in 2008 the Indian space mission to the Moon, Chandrayaan-1, found evidence that the lunar regolith contains water. More specifically, it measured the spectrum of light reflected off the surface and found hydroxyl. And in 2010 the radar experiment on Chandrayaan-1 discovered direct evidence for large quantities of water ice that exist within dark craters at the poles of the Moon, where the temperatures never rise above minus 160 degrees centigrade.

---

[34] Saal, A.E. et al. 2008, "Volatile content of lunar volcanic glasses and the presence of water in the Moon's interior", Nature, 454, 192.

This has since been confirmed by other orbiting lunar satellites, and although this is all indirect evidence, there could be as much as a billion tons of water ice within the lunar craters.

But where did this water come from? One possibility is that hydrogen ions (protons) from the solar wind impact the surface of the Moon and combine with oxygen atoms on minerals to produce water or hydroxyl. Another possibility is that the water has been delivered to the Moon by comets and asteroids, which are both known to contain significant amounts of water. However, the most recent research analysing isotopes of hydrogen from the water trapped in Moon rocks have revealed that the water on the Moon is identical to the water on Earth.[35] This suggests that the water on Earth and the Moon may have a common origin. This is all rather fascinating – particularly because the presence of large quantities of water ice would make the possibility of a lunar base far more feasible. But it also raises a question – how did the water get on the Moon?

It is thought that the water on Earth was delivered via asteroids over a long timescale. The new results tell us that this cosmic delivery of water must have mostly occurred prior to the formation of the Moon – in the first sixty million years after the formation of the Earth – otherwise there would be a greater difference in the isotope ratios. This also suggests that

---

[35] Saal, A.E. et al. 2013, "Hydrogen Isotopes in Lunar Volcanic Glasses and Melt Inclusions Reveal a Carbonaceous Chondrite Heritage", Science, 340, 6138, 1317.

the water on Earth survived whatever cosmic catastrophe led to the formation of our Moon.

If the Moon was once part of the Earth, then how did part of our planet end up there in space, orbiting our world? Now is a good time to look at what we know about the origin of our Moon.

## 4. The Origin of the Moon

*For what is the moon, that it haunts us,*
*this impudent companion immigrated*
*from the system's less fortunate margins,*
*the realm of dust collected in orbs?*

This short excerpt is from the author and poet John Updike's *'Half Moon, Small Cloud'*, published in the 2009 collection 'Endpoint'.[36] In four lines it captures the mysterious nature of our Moon, old ideas about its formation and the ancient views of our solar system. No doubt countless humans must have wondered as to where our spectacular celestial neighbour came from.

In Norse mythology, the Moon was made from the sparks and embers from the primordial fire world Muspelheim. In Aztec mythology, the Sun and Moon were once equally bright and resulted from the sacrifice of two gods – realising that two Suns would overwhelm the world, the remaining gods threw a rabbit at one of the Suns, and it became the Moon. An ancient Chinese creation myth tells the story of the creation of the Moon from the essence of Ying and Yang. In the Qur'an, Allah created the Sun, the Moon, and the planets, each with their own individual paths. In Christian folklore, god made the two great lights – the greater light to rule the day and the lesser light to rule the night. Both these latter stories may have their origin in the Babylonian creation myth *'Enuma Elish'*, which is preserved on over a thousand lines of cuneiform text on seven clay tablets recorded in the 12th century BC. It

---

[36] "Endpoint and other poems", John Updike, Knopf 2009.

describes the creation of the Moon, the Sun and day and night by the all-powerful god Marduk.

These are all very romantic stories and probably the best that anyone could do given the only information available at the time – the extant of our knowledge about the Moon prior to the 16th century was that it was a large spherical object that reflected the light of the Sun. No new insights into the cosmos followed after the ancient Greeks for over one thousand years. Before the second scientific revolution that began between around the 16th century, one had to try to make sense of the celestial objects in the context of religious teachings and dogmas. The first scientific ideas on the origin of the Moon only emerged after the invention of the telescope revealed more information as to its nature.

One of the first attempts to explain the Moon's formation is recorded in Rene Descartes *'Le Monde'* (The World). Descartes completed his major work by 1633 but withheld it from publication because of his fear of the church and its condemnation of the heliocentric ideas and particularly the trial of Galileo in 1633. Because of Galileo's support of the heliocentric model, the Catholic Church found him 'vehemently suspect of heresy' and sentenced him to indefinite imprisonment. By 1649 Descartes had gained a reputation as one of the greatest European philosophers, but he died in 1650. In his 2009 book *'Der rätselhafte Tod des René Descartes'*[37], Philosopher Theodor Ebert argues that Descartes was assassinated by a Catholic priest using the poison arsenic. All of Descartes' works were placed on the Index Liborum

---

[37] "Der rätselhafte Tod des René Descartes", Theodor Ebert, Alibri 2009.

Prohibitorum, the list of books forbidden to be printed or read by Catholics – a list that remained in place until 1966.

In *'Le Monde'*, finally published in 1664, Descartes discussed many topics, from biology to cosmology. Descartes imagined a universe filled with pieces of matter of various sizes, shapes and motions, evolving into a system of numerous vortices rotating around stars. Larger pieces of matter became planets whose orbits around stars are maintained by collisions with smaller pieces of matter, each planet developing its own vortex of smaller bodies, thus forming planets with their systems of moons. Descartes was attempting to describe a generalised picture of moon formation, motivated by the discovery of the moons of Jupiter by Galileo.

Influenced by Rene Descartes and the Swedish philosopher Emanuel Swedenborg, Immanuel Kant and Pierre-Simon Laplace independently proposed a similar theory for how stars and planets form. These are the basis of modern theories for the origin of the solar system. In 1755 Kant argued that gaseous clouds (nebulae) slowly rotate, gradually collapse and flatten due to gravity, eventually forming stars and planets.[38] The theory of Laplace also featured a contracting and cooling proto-solar cloud—the proto-solar nebula. As it cooled and contracted, it flattened and spun more rapidly, shedding a series of gaseous rings of material. The planets condensed from this material and the

---

[38] "Allgemeine Naturgeschichte und Theorie des Himmels", published anonymously by Immanuel Kant, 1755

Moon was envisaged to condense from a ring of matter spun off from the rotating gaseous proto-Earth.

Since the 17th century the leading theory for the origin of our Moon has changed several times. Over the past thirty years a new favoured scenario for the formation of our Moon has developed, yet there is still no generally accepted consensus among the scientific community that this is the reality. Prior to the Apollo program, a successful formation theory had to reproduce the basic observable quantities – the relatively large size of the Moon and its orbit. But there was one more big clue as to the origin of our Moon that was known already before the Apollo mission, and that was the fact that it has a density that is less than the Earth. This knowledge came from the first accurate measurements of the mass of our Moon.

**The mass of the Moon**

The mass of an object is a measure of the number of atoms it is made of, but how can you measure the mass of something so far away that had never been visited?

The first estimate of the mass of the Moon was made in the 17th century by Isaac Newton in a rather ingenious way. Newton knew that the Moon and Sun caused the ocean tides and that the height of the tides depended on the distance and masses of the Moon and Sun. Thanks to the ancient Greeks the distances to the Moon and Sun were already known. Newton could also estimate the mass of the Sun using his theory of gravitation applied to the orbit of the Earth. He could then measure the mass of the Moon, knowing that the ocean tides due to the Moon were about twice as high as those due to the Sun. Newton inferred that the Earth was about forty times as massive as the Moon, from which he could estimate the density of the Moon and stated *"Therefore the body of the Moon is more dense and more Earthy than the Earth itself."*

Newton's estimate was not very accurate, and the main reason was that he used the tidal height in just one location, the estuary of the river Avon in England. However, the height of the tides can vary significantly from place to place because the tides depend on many complicated factors such as the depth of the sea, the boundaries of the surrounding land and the complex ocean currents. By the 19th century, using more care in the measurements, the Earth was found to be closer to eighty times the mass of the Moon. This was later confirmed by another technique to measure the mass of the Moon, which used the parallax of distant stars to determine the distance to the centre of mass of the Earth and Moon. This measurement implied that the Moon was about half as dense as the Earth and it turned Newton's original statement upside down. The Moon was actually less Earthy than the Earth itself!

We now know the mass of the Moon to an accuracy of one part in a billion, thanks to lunar orbiting satellites. Whilst the accuracy of this number may not seem too important, it provides another constraint on the theories for the formation of the Moon. From its mass and size, the average density of the Moon is just 60 percent that of the Earth, close to that measured for large asteroids. This led many scientists to favour the idea that the Moon was a random solar system body captured by Earth's gravitational pull. However, this theory was later abandoned once the composition of the

Moon rocks revealed that the Moon was made of material that is nearly identical to that of the Earth.

At the end of the 19th century a new theory became the accepted idea of how the Moon formed. And although it is a theory that was eventually superseded almost a hundred years later, it contains the essence of how I think the Moon formed. The new ideas came from George Darwin, and we will encounter more of his pioneering work on the history of the Earth and Moon later. Whilst several researchers before Darwin had realised that the Moon would be drifting away from Earth because of the effects of gravitational tides, Darwin was the first to consider tracing the Moon back in time. He reversed the cosmic clock and calculated the history of the Moon, realising that if the Moon was moving away from Earth today, it must have been much closer to Earth in the past.

Darwin's lengthy analyses of the history of the Earth and Moon led him to the conclusion that the Moon must have once been so close to Earth that they were touching, or in other words, the Moon actually originated from Earth itself.[39] At this time the Earth would have been spinning far more rapidly than today, with a rotation period of about five hours long. If the Earth was spinning at that speed it would still hold matter onto its surface. But if it was spinning so fast that the length of the day was about two hours long, then the centrifugal force would be so strong that matter would be flown from its surface. And that is how Darwin envisaged the

---

[39] "On the bodily tides of viscous and semi-elastic spheroids, and on the ocean tides upon a yielding nucleus", 1879 Philosophical Transactions of the Royal Society, 170, 1.

Moon formed, from debris spun-off a rapidly rotating proto-Earth.

At this time, it was not known that there was a giant dense metal core at the centre of our Earth, or that the Moon lacked such a large metallic core. Had this been known, Darwin could have claimed that his model also gave rise to the correct structure of the Moon, since material ejected from the outer mantle of the Earth would not contain much iron. However, support for Darwin's idea came from the geologist Osmond Fisher who in 1889 suggested that the scar left by the Moon's separation from Earth did not completely heal.[40] Fisher suggested that the Moon originated from the regions that became the Pacific and Atlantic Oceans. This proposal came before the knowledge that it was plate tectonics which had separated the continents, so Fisher's idea seemed plausible at the time.

Darwin struggled to find a mechanism that could increase the Earth's rotation to the break up speed. He considered a contracting proto-Earth that rotated faster thanks to the law of conservation of angular momentum (rather like a ballet dancer who spins faster by bringing their arms closer to their body). He also suggested that the gravitational tides from the Sun could spin up the Earth, but he was not completely convinced of either mechanism. Despite this, by the early 20th century, Darwin's theory was widely accepted, even though there was no known way of spinning up the Earth fast enough that it could shed material to form the Moon.

---

[40] "Physics of the Earth's crust", 1889, Osmond Fisher, The Macmillan Company, New York

The theory of continental drift, proposed by the German polar researcher Alfred Wegener in 1912,[41] did not become widely accepted until the 1950's, when numerous independent lines of evidence all supported his idea that the continents slowly drift upon the surface of Earth's hot interior. By this time Darwin's theory of the formation of the Moon began to fall out of favour since the oceanic regions that were supposed to be the sites from where the Moon originated were realised to be due to continental drift.

Other models for the formation of the Moon were proposed. The nebula hypothesis of Kant and Laplace was invoked in more detail by the astronomer Gerard Kuiper who suggested that the Earth and Moon developed as a double planet within the primordial solar-system nebulae.

It was the American scientist Harold Urey in the 1950's who brought fresh ideas into play. Urey promoted the idea that the formation of the Moon occurred separately from the Earth, and that the Moon was subsequently captured by our planet. However, Urey calculated that the chances of such a capture would be remarkably small, unless there was a vast number of such objects in the early solar system. That led Urey to develop new ideas on the formation of the planets, in particular using geochemical evidence to argue that planet formation involved the aggregation of numerous small cold bodies like asteroids.

By the beginning of the 1960's there were three main Moon formation theories. The capture hypothesis, co-accretion

---

[41] "Die Entstehung der Kontinente", 1912, Alfred Wegener, Geologische Rundschau, 3, 276

within the Kant-Laplace proto-nebula model and Darwin's fission hypothesis. In 1958 NASA was founded with the goal of advancing human knowledge of space, and, as we have already learned, Urey played a key role in defining its scientific strategy, making the case for the journey to the Moon to uncover its history and origin. By 1964 four different programs of lunar exploration were being discussed, and in the same year the first conference specifically devoted to the question of the origin of the Moon took place. There followed widespread expectation that the Apollo missions would settle the question of its origin, and this was frequently cited as one of its main scientific goals.

In 1974, shortly after the Apollo program ended, a scientific conference on planetary satellites took place at Cornell University. At this meeting the astronomers William Hartmann and Donald Davis presented a new model for the origin of the Moon – that it formed from the debris created after a giant impact had occurred on Earth.[42] At the same meeting two other astronomers, Andrew Cameron and Bill Ward, announced that they were also developing a similar model.[43] However, at the time little attention was given to these new ideas.

The next conference devoted to the origin of the Moon took place in Kona ten years later in 1984, and brought together all of the results from the Apollo missions. What emerged from this scientific meeting was that all of the pre-Apollo ideas on

---

[42] "Satellite-sized planetesimals and lunar origin", Hartmann, W.K. and Davis, D.R. 1975, Icarus, 24, 504.

[43] "The origin of the Moon", Cameron, A.G.W., and Ward, W.R., 1976, Abstracts of the lunar and planetary science conference, 7, 120.

the formation of the Moon were disfavoured and the giant impact model became the most widely accepted formation scenario.[44] It was a natural extension of the prevailing idea on how planets themselves form.

**From stardust to planets**

The space between the stars in our Galaxy is not completely empty. The interstellar medium is an extremely diffuse gas of mainly hydrogen and helium. It also contains a mixture of atoms and dust that has been synthesised by generations of stars over cosmic time. It is from these ashes of long dead stars that new stars and their planets form. Although there is a great deal still left to learn about the formation of stars and planets, I will describe the general scenario that is emerging from current astrophysics research.

Each year, several new stars appear in our Galaxy. Stirred by the supernova explosions of dying stars, large clouds of gas are constantly forming within the turbulent interstellar medium. Some of these clouds become so massive that they begin to collapse. As gravity pulls everything inwards, any small amount of rotation in the initial cloud is amplified as it shrinks to a smaller volume. The dense central core of the collapsing cloud forms a proto-star, surrounded by a swirling disk of gas and dust. It is within this rotating proto-planetary disk that the planets form. From Earth, we can see these giant disks of gas surrounding newly formed stars, and we can see older stars with planets. But we cannot observe the actual process of planet formation since our telescopes cannot

---

[44] "The giant impact hypothesis: past, present (and future?)", Hartman W.K., 2018, Philosophical Transactions of the Royal Society, 372.

resolve the details. Therefore, the steps from dust to planets is still debated, and there are several unsolved problems.

The basic idea is that within the proto-planetary disk the atoms and molecules begin to collide and stick together. Tiny clumps of dust form, held together by weak intermolecular forces and as fragile as a snowflake. Some of these dusty clumps rapidly outgrow the rest by sweeping up more material as they orbit around their newly formed star. As these clumps become larger, their gravity helps attract and hold onto additional material. Eventually, there are trillions of giant boulder sized objects to mountain size aggregations of dust, ice and rocks.

Some of these objects grow so large that their own gravity is sufficient to pull them into a spherical shape. Thousands of small proto-planets are left orbiting the star. Giant collisions between these objects are common. Some of these encounters are so violent that they shatter each other into pieces, other collisions lead to their coalescence. Finally, several terrestrial planets may have formed, sweeping up most of the smaller objects in their orbital vicinity. Observations and numerical simulations suggest that 'from dust to planets' takes less than 100 million years.

The asteroids are the last remnants of the planetary building blocks and shattered debris from all the impacts between proto-planets during our early solar system. The last truly giant impact on Earth occurred shortly after the Earth had assembled. Around 4.51 billion years ago, a planet the size of Mars is thought to have collided with Earth. This impact would have vaporised a large fraction of the Earth's surface and the smaller planetary object that hit it. The debris that was ejected into orbit formed a spinning disk of material around the Earth. It is from this material that our Moon is thought to have formed, in a similar fashion as the Earth itself. That is the essence of the 1974 giant-impact model for the origin of the Moon.

## Problems of the impact model

Giant impacts between proto-planets are a natural consequence of the late stages of planet formation, when the proto-planets are growing in size by colliding and merging together. However, some scientists believed that the type of collision which led to the formation of our Moon must be very rare – it required another planet-sized body to hit the Earth at a high speed and at just the right angle to create a disk of debris orbiting the Earth. I became interested in this particular problem after it was shown that our Moon stabilises Earth's climate – some astronomers have argued that without our Moon complex life could not arise – a grand solution to the Fermi Paradox[45] if our Moon were rare.

In 2010 our research group at the University of Zurich began working on a new computer code to simulate the late stages of planet formation. Our aim was to calculate the probability of Moon-forming impacts on a proto-Earth. The code ran entirely on the fast graphics cards designed to make numerous simultaneous calculations of 'collisions' between objects in computer games. We used the optimised hardware to calculate the orbits and collisions that occur between proto-

---

[45] The Fermi Paradox is the observation that since most stars host planets, then why haven't we found evidence for life out there?

planets during their formation. Our numerical simulations revealed that around one in ten Earth-like planets would undergo such a collision with the correct energy and geometry to create a massive Moon.[46] That's not particular rare given that there are an estimated 100 billion Earth-like planets in our galaxy alone – it still leaves 10 billion with massive Moons!

But is this the correct scenario by which our Moon formed?

There are four key observations that any theory for the origin of our Moon must explain. Firstly, the mass of our Moon relative to the Earth is rather large compared to the other solar system planets. Secondly, our Moon has a lower average density than the Earth because of its much smaller central iron core. Thirdly, the composition of our Moon is virtually identical to the Earth's. Fourthly, the present-day rotation rate of the Earth and the Earth-Moon distance must be a possible outcome of any successful model. In addition, any model should also be able to explain why the Moon formed when it did, why it was once completely molten and why the lunar surface has water that is identical to Earth's water. Finally, any theory should be plausible and not rely on rare events.

That's quite a list of observations that need to be reproduced by any successful Moon formation theory!

The older 'capture scenario' by which the Moon was somehow captured by the Earth's gravitational field cannot explain why the Moon is identical to the Earth. And it is

---

[46] "How common are Earth-Moon planetary systems?" Elser, S., Moore, B., Stadel, J, Morishima, R. 2011, Icarus, 214, 357.

difficult to envisage a scenario by which such a large solar system body could be captured by the Earth. The nebula co-formation model cannot explain the lower density of the Moon, nor the present day spins of the Earth and Moon.

The giant impact origin of our Moon was thought to explain all of these observations. The large mass of our Moon arises because whatever impacted our Earth was at least as large as the planet Mars – the energy of the collision would then have been sufficient to scatter over a Moon-mass of debris into orbit. The colliding body has sometimes been referred to as Theia, the mother of Selene, the Moon goddess in Greek mythology.

To scatter material from the Earth into orbit, the collision must have been a glancing blow and not a head on crash. This collision geometry and a large impactor would have naturally led to a rapidly spinning Earth and debris with enough angular momentum to match the present day spins of the Earth and Moon together with their current angular momentum.

The lower density of the Moon arises naturally, because the collision would have preferentially sent material from Earth's mantle into space, leaving our iron core intact, while at the same time the iron core of the colliding planet would have sank to the centre of Earth. The similarity of the material within the Moon and Earth arises naturally, since most of the Moon is made from the mantle material of Earth.

However, this similarity turns out to be the greatest problem with the giant impact model. The rocks on Earth and the Moon are just too similar. As we learned in the last chapter, the recent isotope analysis of Moon rocks show that they are identical to the Earth's to a level of two parts per 100,000. That is too close for the giant impact model, since part of the colliding planet also ends up mixed in the debris from Earth that forms the Moon.

Wherever the Mars-sized planet that impacted the Earth formed in our solar system, it would have had significantly different chemical abundances of the various elemental isotopes. For example, the oxygen isotopes in rocks on Mars are different from Earth at a level of three parts in 10,000. This difference in oxygen isotope abundances is due to the fact that Mars formed further away from the Sun, where the initial proto-planetary disk was cooler. Since the Moon forms from a mixture of the impacting planet and the Earth, the Moon rocks should reflect this difference.

It is unlikely that any weathering or subsequent processes could have changed the oxygen isotope fractions, and in 2012 researchers measured the isotope ratios of titanium-47 to titanium-50 and found the same results.[47] Since titanium is very robust against heat, pressure or chemical changes, this confirms that the Moon was indeed once part of the Earth.

Several variants of the giant impact model have tried to explain the fact that the Moon is so similar to Earth. Some researchers have argued that the Moon must have formed with the same composition as the Earth, but others have claimed that this is highly unlikely. Scientists have since proposed that a mission to Venus could help resolve this

---

[47] "The proto-Earth as a significant source of lunar material", Zhang, J. et al., 2012, Nature Geoscience, 5, 251.

problem. Mars is the only planet of which we have analysed rocks, via robotic missions or the few Mars meteorites that have ended up on Earth. Analysing the rocks on Venus would allow us to test the isotopic variation of rocks on another world.

Perhaps the most promising suggestion is that the planet that collided with the Earth was far larger than Mars.[48] Another proposed solution was that the Earth was bombarded with multiple smaller collisions over a period of millions of years, each forming small moonlets in orbit about the Earth.[49] Over time these merged together to form our Moon. In this case each small moonlet was made entirely from the debris from Earth. And in yet another study, it was suggested that a 'hit and run' scenario had taken place.[50] The impacting planet skimmed the surface of the Earth, scattering mantle material into orbit, but never merging with the Earth.

In 2018 it was proposed that a super-giant collision must have taken place, which vaporised a large fraction of the Earth and the colliding object.[51] This would have formed a giant doughnut of debris around what was left of the Earth, and because the collision was so violent the debris from the mixture rained down onto Earth and the forming Moon homogenising their surface compositions. However, such a

---

[48] "Forming a Moon with an Earth-like composition via a giant impact", Canup, R.M. 2012, Science, 338, 1052.

[49] "A multiple-impact origin for the Moon", Rufu, R., Aharonson, O., Perets, H.G. 2017, Nature Geoscience, 10, 89.

[50] "A hit-and-run giant impact scenario", Reufer, A., Meier, M., Benz, W., Wieler, R. 2012, Icarus, 7, 21.

[51] "The origin of the Moon within a terrestrial synestia", Lock, S.J. et al., 2018, Journal of Geophysical Research, 123, 910.

giant collision would have also vapourised the existing water on Earth which we learned was present before the Moon forming impact.

## A cosmic merger

In the past few years, many different scenarios have been proposed. However, none of the existing proposals have gained acceptance amongst the astronomical community. This seems like a solvable problem, and I was intrigued to investigate further. A few years ago my PhD student, Miles Timpe, and I began our own series of numerical simulations to study the origin of the Moon. The first giant impact simulations carried out in the 1980s used computers that were far less powerful than today – we now have access to supercomputers that can treat the collisions with a much greater degree of reality. And because the older computers were slower, only a small region of possible collision geometries could be studied. Using one of the world's most powerful supercomputers located at the Swiss National Supercomputing Centre, we have recently carried out over ten thousand high resolution impact studies – by far the largest such study ever performed.

The number of variables is large – the size and composition of the proto-Earth and impactor as well as the speed and angle of the collision. To further complicate things, the proto-Earth and impacting planet may have already been spinning, and that also changes the collision outcome. Even with today's computers the parameter space is very large. However, there seems to be a large region of possible collisions that was previously unexplored and can reproduce all of the observed constraints.

One possible scenario that we have found is that the collision was not a giant impact, but rather a merger of two similar sized planets. In this case the two proto-planets would have each had about half of the mass of the Earth. The merger proceeds by the two planets spiralling around each other, their gravitational force so strong that they are distorted into ellipsoidal shapes. This form of high-angular momentum merger had not been previously studied since it was thought the proto-planets would pass each other unscathed. However, the gravitational tidal forces during the first near passage of the planets cause both planets to start spinning – this energy comes from the orbital energy, which results in the planets spiralling closer together until they merge into one rapidly spinning Earth-like planet.

The new world – our Earth – is molten hot from energy of the collision. Its surface temperature is over one thousand degrees, and its radiant glow would be visible to any alien observers watching the formation of our solar system from a nearby star-system. It takes thousands of years before its surface solidifies to form the first rocky crust.

The material from both planets completely mixes, their iron cores sinking together to form the centre of the new planet. The angular momentum from the orbital motion is completely transformed into the rotation of the merged spinning planet. This causes the merger remnant to rotate at a speed close to the break-up speed. The rapidly spinning planet is distorted into the shape of a rugby ball, rotating once every few hours. Following the merger, several Moon-masses of material is flung outwards into space, leaving symmetrically from either end of the spinning ellipsoid. This is reminiscent of Darwin's original model for the origin of the Moon.

After just a few Earth days, an extended disk of material forms around the spinning Earth – more than enough material from which our Moon can form. Even if the two merging planets had different compositions and isotope ratios, the final mixing of the material is complete – the material in the mantle of the final planet is identical to within a fraction of a percent of the material in the lunar forming disk. Such a scenario explains all of the key observational constraints described earlier.

The next step is to run the simulations forwards for a timespan of years rather than days to follow the formation of the Moon within the ejected material. We may still have a long way to go before the formation theory for our Moon is mature and has been fully developed. And to test our ideas on how the Earth and Moon formed may require a broader exploration of our Moon. It will be interesting to see what the future holds. But before we explore the future of our Moon and the reasons why we should return, let's go back to the beginning and look at the development of all the knowledge, ideas and influences, of our Moon.

## 5. The First Astronomer

Today, more than half the world's population live within brightly lit cities and urban areas. From these light polluted vantage points often just a few dozen stars can be seen. I am writing this text in Davos, a reasonably dark location in Switzerland, but my astronomical photographs suffer from the light of Milan 150 kilometres away. It is a great shame that most people have never witnessed the sight of a truly dark night's sky with its thousands of sparkling colourful stars.

I was fortunate to spend time at the Paranal Observatory high in the Atacama Desert of Northern Chile and will never forget when I saw the night's sky from this truly dark location. It was breathtakingly beautiful. The stars in the Milky Way cast shadows, and our neighbouring small galaxies, the Magellanic Clouds, were like swirls of glitter nestled amongst the constellations. When the Moon rose it seemed like daylight was beginning because it was so bright, and for a moment I was shocked – it was upside down from how I was used to seeing it in the northern hemisphere. I quickly realised the obvious reason why that was the case, but it's nice to see first-hand proof of our spherical world.

I have often wondered how it must have been to have lived on a planet with no chemical pollution in the air and no light pollution in the sky. Our ancestors did not have all the technology or the medical advancements we have today. Yet I could not help feeling jealous of those who lived under such dark and clear skies. It was easy to see why our ancestors thought that the gods lived up there amongst the stars. It made sense that connections were made with the constellations, and why the Moon became such an important role in mythology and everyday life.

So how did the path towards our knowledge of the Moon begin? Let's imagine the attempts to make sense of the world by the first Neolithic astronomer. Let's call her Eve and let's start at the beginning and see what Eve could learn by eye, by regularly observing the sky and watching the motions of the Sun and Moon, planets and stars over the course of a year.

The most obvious cyclical patterns known to all would be the seasons. Be it the first fall of snow in the distant mountains, the seasonal monsoons and floods that gave rise to the fertile valleys, the spring blossom and winter snows. Perhaps most importantly, as our ancestors transitioned from foragers and nomads to begin settlements and agriculture, it would become important to know when to plant crops. The seasons were the most important cycle, but the times of these cyclical events would appear to vary and depend on the weather. So how could Eve devise a means of keeping track of time beyond the regular occurrence of night and day?

The motion of the Sun and Moon across the sky and the clockwork-like cycle of the lunar phases are the most obvious celestial phenomenon. For time periods longer than a day the Moon would seem to be a good candidate for the first celestial clock.

Over a lunar month, the Moon cycles through its phases, from crescent Moon to full Moon and back. Eve would notice that as the Moon moves closer to the Sun it transforms towards a new Moon and then disappears for several nights.

Eve would also see that halfway through this cycle, when the Moon is half illuminated, it is close to overhead at sunset and sunrise. A careful observer of the skies like Eve would observe that a full Moon always happens when the Moon is nearly opposite the Sun. With clear and dark skies perhaps she would notice the faint glow of Earthshine from the dark part of the half Moon.

It would have been an extraordinary feat of intellect, equivalent to that of Isaac Newton, if Eve were able to put all this together to ascertain that the Moon received the light of the Sun, its phases were due to its spherical shape, and its fainter glow was the reflection of light from the Earth. Perhaps Eve simply came up with a story to explain these patterns of events. All this chasing around of the Moon by the Sun led Eve to her first myth, perhaps a tale of birth, death and rebirth – that the Sun gives birth to the Moon after which it grows to its brightest before the Sun chases it to destroy it.

The complete cycle of phases of the Moon from new Moon to waning crescent and full Moon and back to waxing crescent and new Moon take place over what is called a *synodic month*. The Moon shines from the light of the Sun and the different phases are caused by the angle between Earth, the Moon and the Sun. The synodic month is the time it takes for the Moon to rotate around the Earth and back into the same position relative to the Sun. The light of a crescent Moon is curved because the Moon is spherical. You can demonstrate this by holding a white ball in front of a bright light and looking at it from different angles. This might seem trivial, but this explanation of the lunar phases was not given until Aristotle in the 3rd century BC. Perhaps that was because white balls and bright lights were not available until late in human history!

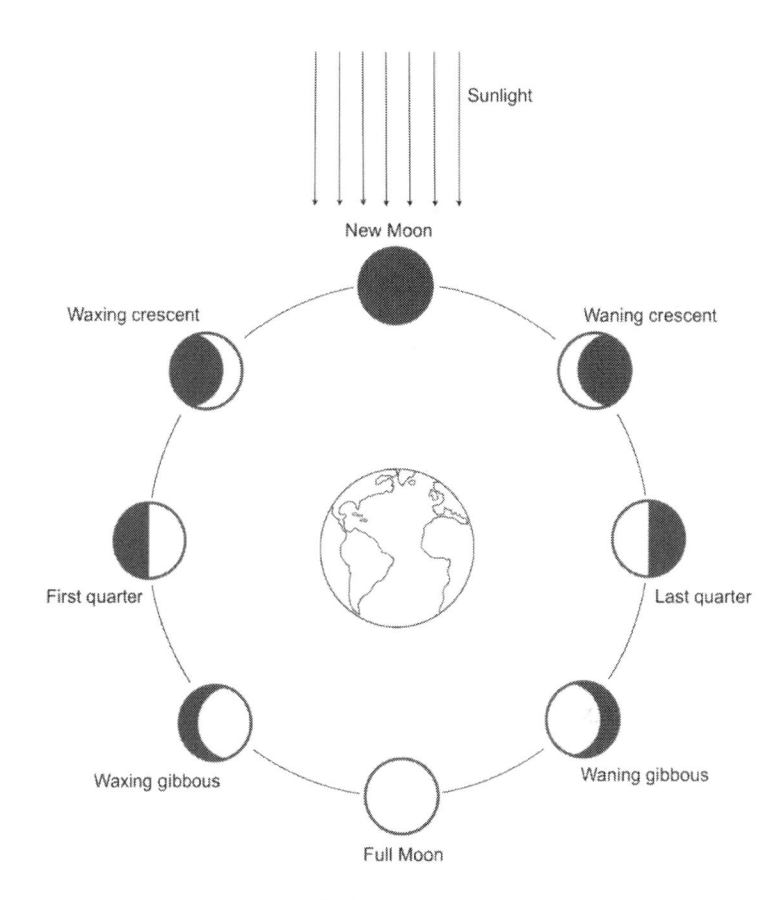

*The lunar phases*

Eve would soon notice that the cycles of the Moon repeated about every 29 or 30 days. That would seem like a convenient regular pattern with which to divide up the year of seasons making it easier than counting all those days of the year. But Eve would find it difficult to accurately determine the length of the monthly lunar cycle because the Moon appears very close to full for three nights. Frustrated with guessing the precise day that hosted the fullest Moon, Eve might look for the first sight of the Moon after a new Moon – an easier way to record the start and end of the lunar cycle. After making observations over several lunar months it would become apparent that the lunar cycle took 29 and a half days. And Eve would notice that after about 12 of these lunar cycles, around 354 days, the seasons would repeat.

So Eve constructed the first calendar defining 12 lunar months as a year, great progress! Almost all early civilisations began their timekeeping with a lunar calendar. But all was not well. Eve was a persistent observer of the skies, and she noticed after a few years that the seasons were not synchronised with the lunar year she had defined. Winter and summer were beginning later and later with respect to her 12 lunar months. Something was wrong.

**Equinoxes and solstices**

Eve might have then shifted her attention to the Sun since the Sun was more clearly connected with the seasons. It would have been apparent that over the course of the year, the position of sunrise and sunset slowly move northwards when the climate is getting hotter and southwards when winter approaches. And that from winter to summer the days slowly become longer and the Sun rises higher in the sky. This is because of the motion of our tilted Earth around the Sun, but this would not be known to Eve.

Eve learned to keep track of the sunrise and sunset positions, perhaps with distant landmarks or by aligning sighting stones on the ground. Only after about 12 lunar months and 11 days would the Sun begin to rise and set in the same places, and that time period coincided with the seasons. It would be clear to Eve that the Sun defined the seasons but the Moon was a more convenient way of dividing the year into countable parts. A more accurate way of measuring the year was needed – but that would prove to be quite difficult. Of course the seasons defined the year, but people needed to have a measure of the time that was independent of the weather or the first flowers. Most crops would be planted well before the trees began to blossom. So how would you determine that the Earth has made exactly one complete orbit of our star?!

The Sun rises exactly in the east and sets exactly in the west only twice a year, and these days are marked by close to 12 hours of sunlight and 12 hours of darkness. These are the equinoxes, and they occur in March and September in today's calendar. The equinoxes occur when the equator of the Earth is precisely aligned with its orbit about the Sun, the ecliptic plane.

The day of the northern-most extreme points of sunrise and sunset is the longest day in the northern hemisphere and is called the summer solstice. On this day the Sun reaches its highest point in the sky. Similarly, the day of southern-most sunrise and sunset is the shortest day in the northern hemisphere, and is called winter solstice. On this day the maximum height of the Sun is the lowest of the entire year.

The solstices are separated by about 183 days, dividing the year in half, and the equinoxes occur about 91 days later.

If any of the equinoxes or solstices could be measured, then that would be a good way of noting a year has passed, and the lunar calendar could be resynchronised with the solar year.[52]

There are various ways how Eve could determine the dates of the equinoxes and solstices. But it's not that easy. The precise day of the turnaround points for sunrise and sunset would be difficult to determine. The position of the rising Sun on the eastern horizon changes by just a tenth of a degree over the days around the solstices. The shortest shadow when the Sun is overhead (passing the meridian) from a one metre stick during the days around the solstices changes by just a few millimetres. And the length of day and night around the equinoxes change by just a couple of minutes a day.

But Eve was persistent and carefully laid markers of rocks on the ground and adjusted their positions each year so that the shadows and sunrises of these key dates could be accurately measured. Eve would also notice that just before sunrise or just after sunset the same constellations would be rising above the horizon on these dates. That would give her another means to keep track of these cycles. Only by taking observations over decades could Eve pinpoint the precise days of the solstices and equinoxes. With patience, Eve eventually measured that the year was about 365 days long.

---

[52] "The history of the tropical year", 1992, J. Meeus and D. Savoie, Journal of the British Astronomical Association, 102, 1, 40.

## A cosmic calendar

With this knowledge Eve could correct the lunar calendar to match the solar calendar, perhaps by simply adding an extra eleven days each year or an extra lunar month every couple of years – a so called intercalary month. Eve had now developed the lunar-solar calendar and could correctly announce the start and end of the seasons, when crops should be planted and harvested, when to watch out for floods or snow and perhaps most importantly, when the main parties to celebrate the yearly cycles of life should occur!

Now we have reached a stage reached by most early civilisations. Before the times when global trade routes spread knowledge, the lunar-solar calendar was independently developed across our planet, from the Inuit to the Aborigines. At this point Eve would have been considered to be quite a mystic or sage to her society, and she would be employed as a full-time astronomer to keep track of the calendar. Monuments would be built to align with the celestial events and as places to congregate. The first astronomers would take their rightful place amongst the upper echelons of primitive societies!

The earliest Egyptian calendar was also based on the Moon's cycles. But later the Egyptians developed another method for keeping track of the year. Viewed from the latitude of Egypt, the 'Dog Star' in Canis Major, which we call Sirius and is the brightest star, rises just above the horizon at dawn every 365 days at about the time when the important flooding of the Nile occurred.

Based on this knowledge, the Egyptians devised a 365 day calendar that seems to have begun around 3000BC.[53] This allowed the Egyptian priests to count the number of days in a solar year to predict when the Nile would flood – this magic-like talent promoted the priests to powerful members of the Egyptian society. The Egyptians also tracked the positions of constellations of stars and linked crop planting and harvesting with the times when they appeared. They used 365 days as one year. They set the month equal to thirty days and adjusted the twelve months of the lunar year by adding five extra days at the end of the year.

But there is a problem here. The priests cannot have failed to notice that every four years Sirius appears one day later. The reason is that the solar year is closer to 365 days and 6 hours. The Egyptians make no adjustment for this, with the result that their calendar slid backwards through the seasons – just like a lunar calendar but much more slowly. The 12 month lunar calendar would resynchronise with the seasons about every 33 years, but it takes 1460 years before Sirius rises again on the first day of the first month. It is known from ancient texts that in AD 139 Sirius rose on the first day of the first Egyptian month. This makes it likely that the Egyptian calendar was introduced one or two full cycles (1460 or 2920 years) earlier, either in 1321BC or 2781BC.

Just a couple of years ago, a colleague of mine at the University of Zurich, archaeologist Michael Habicht, showed me an ancient Egyptian ointment jar from the British museum dating from the Old Kingdom in the 3rd millennia BC. No one

---

[53] "Ancient Egyptian Astronomy", 1974, R.A. Parker, Philosophical Transactions of the Royal Society A, 276, 1257.

had previously noticed that the hieroglyphics mention the heliacal rising of Sirius (the time of year that Sirius rises above the horizon just before the Sun). This confirms the older date as the start of the Egyptian calendar. And reconstructions of the positions of the stars from this period allowed a recalibration of the chronology of ancient Egypt.[54]

Most ancient civilisations kept track of time using twelve lunar months, and an extra month was inserted from time to time to keep in step with the solar year. This happened in Mesopotamia and in republican Rome, and it remains the case today in the Jewish calendar.

The Sumerians added an extra month in every second or third calendar year to keep it in step with the seasons. On one Sumerian tablet dating from around 1800BC, Hammurabi, the most notable King of Old Babylonia, issued a decree: *"As the year is ending too soon, let the month following Ululu be called the second Ululu, but let the taxes due in Babylon in Tashritu be paid in the second Ululu".*[55] We also learn from this ancient text that taxation is as old as astronomy!

The Islamic calendar is still based on 12 lunar months of alternating 29 and 30 days. That gives 354 days in a year and a calendar that bears no relation to the seasons since it constantly slips behind. The regulation of the lunar calendar was one of the most important and at the same time one of the most difficult problems for Muslim astronomers. In Islamic

---

[54] "A new astronomically based chronological model for the Egyptian old kingdom", 2017, R. Gautschy, et al., Journal of Egyptian History, 10, 69.
[55] "Babylonian Astronomy", 1977, W.M. Oneil, The Journal of the Sydney University Arts Assocation, vol 11.

sacred law, the lunar month starts with the first sighting of the new crescent Moon. Depending on the time that has elapsed since the new Moon, this first crescent can be rather tricky to observe, owing to the short time it stays above the horizon and its faint glow compared to the brightness of the western horizon after sunset. Therefore, one needs to know exactly when and where to look to define the start of the month.

Traditionally, observers were sent to locations with an open horizon to report on the visibility of the Moon. If they failed to see it, they would need to repeat their observations the next evening. If the horizon was cloudy, a fixed number of days had to be assumed for the month in question. These uncertainties, especially for the holy month of Ramadan, were problematic. Early Muslim astronomers developed procedures to define the visibility of the new crescent by calculating the difference in the rising and setting times of the Moon and Sun. Over time they developed more complex methods, taking into account the angular separation of the Sun and the Moon and the apparent speed of the Moon.

**The constellations**

Prior to widespread street lighting in the 19th century, the night's sky was so dark that from anywhere on our planet several thousand stars could be seen during a cloudless night. By observing the starry sky each night Eve would have also noticed that five of the brightest stars seemed to move slowly with respect to neighbouring stars. And they all moved across the sky in the same general path – that is the ecliptic, the orbital plane of the planets around the Sun. Most of the time these bright 'stars' would move in the same direction as the Moon and Sun, but sometimes they paused and moved backwards, or retrograde, with respect to the nearby stars. These bright strangely moving stars would later be called planets, after the Greek word 'planetos' which means 'wanderer'.

The five planets visible with the naked eye are Mercury, Venus, Mars, Jupiter and Saturn. The only stars that appear brighter than any of these planets are Sirius and Canopus, which are brighter than Saturn. However, Canopus is not visible from Greece but could be observed by the Egyptians. These strange wandering bright planets, together with the Sun and Moon, took on a special role in ancient mythology as the homes of the prominent gods. Eve would have noticed that two of the brightest planets were usually seen at sunset and sunrise near to where the Sun was. These are Mercury and Venus, and it was the Greek astronomer Aristarchus, and later Copernicus, who realised that these planets were orbiting closest to the Sun.

After several months of following the Moon's position against the stars Eve would have noticed that the same constellations of stars were rising earlier and new constellations were beginning to appear. Each day the Earth has moved just under one degree further around the Sun, so the view of the stars at the same time of night is slightly different. Only after one year do the same constellations rise at the same place and time. The position of certain constellations would be linked with the seasons, such as Orion rising at sunset would mean the onset of winter in the Northern hemisphere. In ancient Egypt, the stars of Orion represented the god Osiris, who was the god of rebirth and the afterlife.

The next step would be to look a little more carefully at the Moon's motion. Eve would notice that its path across the sky was similar to that taken by the Sun and planets, and that each night at the same time, the Moon would be moving past a different group of memorable bright stars. After about 27 days Eve would notice that the Moon was back in the same position with respect to the stars.

Now Eve had discovered the *sidereal* period of the Moon, although she would not yet understand why this cyclic period is about two days shorter than the synodic lunar month. What Eve had measured is the time taken for the Moon to orbit the Earth exactly once. Because of its motion around our rotating Earth, the Moon rises above the horizon about fifty minutes later each day. The sidereal month is shorter than the synodic month, since for the Sun to illuminate the Moon by the same amount we have to wait a little longer. That's because the Earth and Moon are moving together around the Sun, and the Moon has to move a little further around the Earth to receive the same illumination as viewed from Earth. Because we only see one side of the Moon, it must be rotating exactly once each sidereal month.

The length of the lunar day is the time between the Sun appearing in the same place as viewed from the Moon. This must therefore be the same time as the cycle of phases of the Moon – the synodic month. So for an astronaut standing on the Moon, a Moon-day lasts 29.5 Earth days – that's about two weeks of daylight followed by two weeks of darkness.

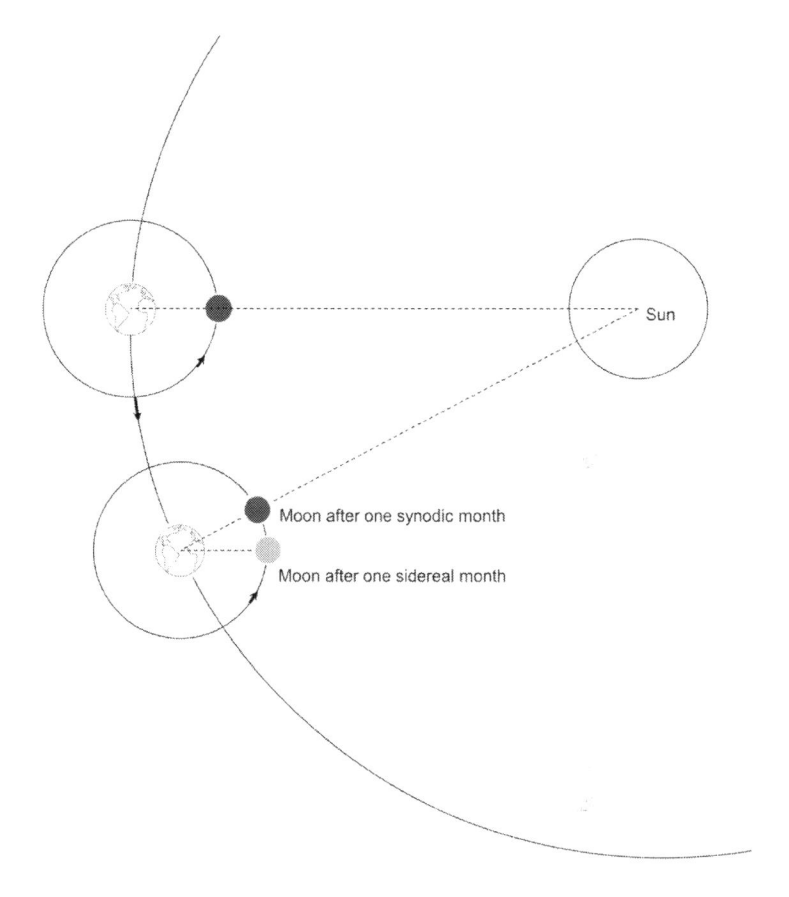

*The sidereal and synodic months*

To keep track of this new cycle Eve divided the regions of the night's sky that the Moon visits into recognisable groups of stars, the first zodiacal constellations. Whilst staring at the constellations each night Eve connected the points of light for fun, creating all sorts of patterns and associations with familiar sights and creatures on Earth. To celebrate this finding Eve created the first cosmic art, painting her favourite constellations onto surrounding walls of rock. All early civilisations came up with their own constellations and their associated symbolism.[56] Some are similar, others very different – the Babylonians divided the path of the Moon along the ecliptic into 12 constellations, the Chinese divided it into 28 lunar mansions.

**Confusion**

Sometimes the stars would appear to fall from the sky and streak across the horizon. There would be rare occasions, about once every three years, when the Moon almost disappeared, turning a dark reddish colour for an hour or so, before lighting up as a thin crescent and returning to a full Moon over the course of another hour. Eve would probably not realise that this was because the Moon passed through the shadow of the Earth – a lunar eclipse – but she would observe that such events always occurred when the Moon was a full Moon. And she had heard tales passed down from her ancestors about a time when the Sun disappeared for a short period during the day. Eclipses of the Sun by the Moon only happen at the same on average after several centuries. However, it was the systematic records of eclipse observations over generations of observers that regular patterns in their occurrence times could be uncovered.

The cyclic periods of the shape of the Moon, the motions of the stars, the strange wandering planets, the eclipses and shooting stars were all rather beautiful and mysterious. But

---

[56] "Origins of the ancient constellations: I. The Mesopotamian traditions", 1998, J.H.Rogers, Journal of the British Astronomical Association, 108, 1, 9.

what did it all mean, and how could sense be made of all of this?

To our first astronomers it would seem that our Earth was somehow a flat stationary object that the Sun, Moon and stars were all moving around at different speeds. The alternative of living on a spinning spherical world that orbits the Sun would make no sense, because surely we would notice the effects of standing on a spinning, orbiting thing?! That was the argument made against a spinning, moving Earth by many early philosophers. But in fact it's rather like travelling at a constant speed in an aeroplane that is turning very slowly, we can't tell we are moving or turning unless we look out of the window at some fixed reference point.

The first astronomers would have tried to make sense of the cosmos, and they would have failed miserably. Without all the knowledge we have now, I'm sure that I would have failed miserably too.

The entire collection of astronomical observations that were available over two millennia ago reflected the fact that we are standing on a spinning spherical world that was tilted with respect to its elliptical orbit about the Sun; that the Moon orbits the Earth tilted to Earth's equator on an elliptical orbit that precesses and wobbles; that the planets orbit the Sun at different distances and their orbital times increase with distance from the Sun and so on. All of these physical effects led to confusion because they were observed as artefacts in the calendar and artefacts in the perceived regularity of the motions of the celestial bodies. And for several millennia no sense could be made of them.

The confusion in the interpretation of these observations is reflected in the stories, myths and folklore that have been passed down through the generations from the earliest societies on Earth. Our brains evolved to make sense of our experiences on Earth, and it is difficult to use those experiences to make sense of the cosmos beyond. Even today, 10,000 years since the Mesolithic hunter gatherers in Scotland constructed the first lunar calendar, we struggle to make sense of the cosmos beyond our solar system. We can't comprehend its size and age or the vast numbers of galaxies, stars and planets it contains.

But thanks to the first philosophers, science did make sense of our world and our place in the solar system. And a large part of that discovery was thanks to the presence of our Moon. But the path to that understanding was long and filled with misconceptions and distractions. Since simple experience could not explain the motions and behaviour of the celestial objects, was it the work of some higher god-like beings? That is certainly a way of obfuscating the answers. Rather like today, we still do not have a clue how our universe began or why there is something rather than nothing, so we may as well just say it was the work of a god. Or we can strive towards a scientific understanding, step by slow step, and that may take another 10,000 years.

The steps that led to humans reaching the Moon, and to our understanding of its origins have also been slow. But retracing these steps tells us about the development of scientific thought and about our own human nature. Telling the story of Eve is how I imagined the first astronomers began to make sense of the cosmos, now it is time to look at the astronomical achievements of the earliest civilisations, and how they tried to make sense of our Moon.

## 6. Our Ominous Moon

Systematic record keeping began with the Sumerians who developed the first system for recording language around 3000BC – the rather beautiful cuneiform script scratched on clay tablets that were dried by the Sun or baked in ovens. These had the advantage that they last far longer than the papyrus scrolls used by the Egyptians or the bamboo used by the Chinese. The Sumer people settled along the fertile river valleys of the Tigris and Euphrates around the 5th millennium, between 5000-4000BC.

In addition to writing, the Sumerians devised a counting system that allowed them to express both very large and very small numbers. They also developed some algebra and geometry, and it was Sumerian mathematicians who first divided circles into 360 degrees and each degree into 60 minutes of arc. The use of 60 as a base of a mathematical system was a good choice: 60 is a number that has many divisors, which makes it easier to deal with calculations involving fractions. This system was later adopted by other cultures and partly continues to this day in our measurement of time, angles and geographic coordinates.

The Sumerians made many other contributions to civilisation that are still part of our everyday life, such as the wheel and crop irrigation. Remarkably, their existence was only discovered in the 19th century as archaeologists searched Mesopotamia for Assyrian and Babylonian relics. Their largest city, Uruk, had ten kilometres of defensive walls and a population of between 50-80,000 people, making it the largest city in the world in 2900BC. Uruk was famous as the capital city of Gilgamesh, hero of the '*Epic of Gilgamesh*', perhaps the first epic work of literature from 2100BC. It is also believed that Uruk is the biblical Erech mentioned in Genesis 10:10, the second city founded by Nimrod in Shinar. Some of the famous stories of the Old Testament, such as Noah's ark and the great flood, may have originated from the Epic of Gilgamesh.

Sumerians believed that the universe consisted of a flat disk enclosed by a dome that was surrounded by a primordial sea. The Sumerian afterlife involved a descent into a gloomy netherworld to spend eternity in a wretched existence as a Gidim (ghost). They regarded the sky as a series of domes made of precious stone that were the homes of their gods. Sin, or Nanna, the god of the Moon, was considered the god of fertility.

For his PhD thesis at the University of Pennsylvania, Mark Hall studied all of the available texts related to the Sumerian Moon god Sin and concluded: *"the Moon-god derived his essential characteristics from the cyclical phases of the Moon. The Moon's ever-renewing cycle was interpreted by the Sumerians as a sign of the Moon-god's inherent power to regenerate himself each month. It was believed that he could bestow this power upon all living creatures. Hence, the Moon-god was a fertility god. His cult was designed to transfer the Moon's generative power to the sphere of man, thus insuring the continued procreation of crops, animals, and the generations of mankind."*[57]

---

[57] "A study of The Sumerian Moon-God, Nanna/Seun", 1985, M. G. Hall, University of Pennsylvania PhD dissertation.

The Sumerian ziggurats were temples constructed for the priests to worship the gods that lived amongst the celestial bodies. They identified the planets Mercury, Venus, Mars, Jupiter and Saturn. The Sumerians used a calendar of 12 lunar months and aligned it with the seasons using intercalated months. What little we know about their astronomy comes from the Babylonians, who copied and refined their techniques.

Around 2200BC there was a long drought across the northern hemisphere that is thought to have devastated many of the emerging civilisations, including the Old Kingdom in Egypt, and those in the Indus Valley and the Yangtze River Valley. This drought lasted over a century and may have been a result of shifts in ocean and atmospheric circulation. The centre of power in the near East shifted south to the city of Babylon, initially a small provincial town that grew into a major capital city by 1800BC.

**Omens in the sky**

The Babylonians took over the techniques started by the Sumerians, perfecting them and creating vast libraries of information of astronomical events. Babylonian priests observed the path of the Moon, stars and planets in great detail and used these recorded movements and information to predict events, which they referred to as omens. This was not just a case of recording details every other day or week, but consistent daily, monthly and yearly records.[58]

---

[58] "Astronomy the Babylonian Way", 2012, J. Anderson, Journal of the Royal Astronomical Society of Canada, 106, 3, 108.

In nearly all early cultures, the eclipses of the Moon were seen as omens of bad things to come. These dramatic reoccurring celestial events were usually associated with death and disaster. The ability to forecast the dates of eclipses meant the Babylonian astronomers were influential advisors to the kings who sought their advice to help predict future events. The eclipses were seen as such important signs from the gods that during a predicted time when an eclipse would occur, the king would put in place a temporary replacement king who would suffer the consequences of any bad omens!

From over a million Sumerian and Babylonian tablets that have been discovered, about a half have been translated, and less than one percent of the tablets discuss astronomy. But that is a comparatively huge record compared with any other civilisation on Earth at this time. It's about 5000 tablets of which 3000 have been studied and published.

Hundreds of years of Babylonian observations of celestial phenomena are recorded in the series of cuneiform script tablets known as the *'Enūma Anu Enlil'*. These 70 tablets record celestial motions, the phases of the Moon, lunar and solar eclipses and weather phenomena spanning centuries. It includes the 'Venus Tablet of Ammi-Saduqa', which was compiled during the 17th century BC and details the times of the first and last risings of the planet Venus over two decades. Much of the content of these tablets is concerned with divination and omens, aiming to interpret the astronomical observations in terms of the fortunes of the king and his empire. Astronomy in Babylonian times was really just a branch of astrology.

The first 13 tablets deal with the first appearances of the Moon on various days of the month, its relation to planets and stars and such phenomena as lunar haloes and crowns. Here is a typical report dealing with the first appearance of the Moon on the first day of the month: *"If the moon becomes visible on the first day: reliable speech; the land will be happy. If the day reaches its normal length: a reign of long days. If the moon at its appearance wears a crown: the king will reach the highest rank."*[59]

Tablet 14 details a basic mathematical scheme for predicting the visibility of the Moon. Tablets 15 to 22 are dedicated to lunar eclipses. The Babylonians used many forms of encoding, such as the date, watches of the night and quadrants of the Moon, to predict which regions and cities the eclipse was believed to affect. Tablets 23 to 29 deal with the appearances of the Sun, its colour, markings and its relation to clouds and storms when it rises. Solar eclipses are discussed in tablets 30 to 39. Tablets 40 to 49 concern weather phenomena, storms, thunder and earthquakes. The final 20 tablets are dedicated to the stars and planets, and use a form of encoding in which the names of the planets are replaced by the names of fixed stars and constellations. Tablet 63 is the famous 'Venus tablet', which reveals the cyclical nature of the orbit of Venus.

---

[59] "Astrological reports to Assyrian kings", 1992, H. Hunger, State archives of Assyria, vol 8.

*The Venus Tablet of Ammisaduqa*

The Babylonian astronomers had laid the foundations of what would eventually become western astrology. The astrology they developed was not the horoscope for individuals based on their birth date – that originated from the Roman era. Rather, they used their ability to predict the motions of the planets and eclipses to try to predict the weather or to provide omens relevant for their king or community.

The MUL.APIN tablets are another important sequence of astronomical records thought to have been made around 1300BC. They contain catalogues of stars and constellations as well as schemes for predicting heliacal risings and the settings of the planets, lengths of daylight at the solstices and equinoxes, the lengths of gnomon (sundial) shadows at different times of day, and intercalations for fixing the lunar-solar calendar. They also show what stars are rising and which constellations contain the full Moon on the solstices and equinoxes in order to judge the disparity of the lunar and solar cycles. They list the stars on the path of the Moon, the major constellations close to the ecliptic, which includes all the Babylonian forerunners to the zodiacal constellations. And, of course, a large selection of astrological omens.

In the later Babylonian Empire, the priests responsible for studying the skies became known as Chaldeans. Although these priest-astronomers continued to use the old star catalogues to predict celestial phenomena, they also developed simple mathematical models that allowed them to predict astronomical events without consulting the records.[60] This ability to foretell the future must have seemed magical to the general population, and the Chaldeans enjoyed positions of great power and respect within the social and religious hierarchy.

The Chaldean astronomers observed that the Sun's motion along the ecliptic was not uniform, though they were unaware of why this was. That knowledge had to wait two thousand years for Johannes Kepler to show that this is due to the Earth moving in an elliptic orbit around the Sun, with the Earth moving faster when it is nearer to the Sun at perihelion and moving slower when it is farther away at aphelion. By the 5th century BC the Babylonians had developed a celestial coordinate system in the form of the zodiac: 12 constellations each spanning about 30 degrees on the sky. Of the Greek zodiacal names we preserve, ten were early Mesopotamian in origin, and perhaps the other two were late Babylonian.

The ancient Babylonians developed sophisticated water clocks that were used to record the position of the Moon and stars to an accuracy of a few minutes. Through their records of lunar eclipses they began to see patterns that could be used to predict future eclipses. One tablet, written around 276BC, records lunar eclipses for a period of 95 years, spanning the reigns of many Kings. The eclipses are arranged in a series of 38 with 223 lunar months between the series – the eighteen year Saros cycle! Each series consists of five sub-series of

---

[60] "Ancient Babylonian astronomers calculated Jupiter's position from the area under a time-velocity graph", 2016, M. Ossendrijver, Science, 351, 6272, 382.

seven or eight eclipses that are six lunar months apart. However, the tablet cannot be a record of eclipses observed at Babylon, since about half the listed eclipses occurred below the horizon or during the day. Also, lunar eclipses are not always total eclipses of the Moon by the Earth, some are partial (penumbral), and there is no record of observed eclipses of this type – as expected since they are barely visible with the naked eye. The tablet is not a record of observations but a prediction of possible eclipses for the coming century!

Another tablet reveals that by the 2$^{nd}$ century BC the Babylonians had made accurate measurements of the synodic, draconian and anomalistic lunar months. In the next chapter I will describe in more detail what these months are, since they are important for predicting when eclipses will occur. They also measured the anomalistic year, which is 365.26 days. It is a measure of the time taken for the Earth to travel from perihelion to perihelion – the point in its elliptical orbit that is closest to the Sun. It is about 20 minutes longer than the tropical year that defines the seasons because of the precession of the Earth – a wobbling motion of our spinning planet. The Babylonians didn't know anything about the Earth's orbit at all. What they observed was the varying speed of motion of the Sun, Moon and stars that circled across the sky.

These were remarkable observations indeed, and the values were known to the Greek astronomer Hipparchus, who adopted them in 150BC to discover the precession of the Earth. The Babylonians had all the records and data needed to infer that the Earth is a spherical wobbling planet, that the Moon orbited the Earth and that the planets orbited the Sun on elliptical orbits. However, Babylonian astronomy was descriptive and predictive and attempted no explanations in terms of physical nature or motions in geometric terms. Astronomy was carried out by priests and primarily for the sake of astrological prediction.

**Astronomy in ancient India**

The Tigris and Euphrates were not the only river valleys around which great civilisations were emerging. After the end of the ice age 12,000 years ago, settlements and agriculture began to appear independently in many regions around the world. In particular, the Indus valley in India and Pakistan and the Yangtze and Yellow rivers in China were home to rapidly advancing societies in the Neolithic period.

Western writers covering topics of the history of science are often criticised for neglecting the contributions from the East. But these developments are not as well documented and researched as in the West, and it is difficult to give due credit. Many historians of science have also written that most of the findings from the East travelled there from Babylonia.[61] But more recent studies have found that many of the results must have been made independently. Many innovations also spread from East to West, such as the base ten counting system and the concept of zero. So, in the same period that

---

[61] "The recovery of Early Greek Astronomy from India", D. Pingree, Journal for the History of Astronomy, 7, 109.

spans from the Sumerians to the Babylonians, what astronomical knowledge was developed further afield?[62]

It is very difficult to disentangle the history of astronomy from the Indus Valley civilisation. They did not leave behind a written record, so it is very difficult to date the ancient texts. The Vedas were not text as such, but a series of rhyming verses that were passed down orally through the generations. They were not written down and translated until two thousand years after they are thought to have been finalised. Various studies reveal the accuracy of these scriptures and that's quite remarkable. Tens of thousands of verses were passed down by memory alone through a hundred generations. It can be difficult to interpret these texts. The rhyming verses use a symbolic language with double meanings. For example, instead of one, the text uses a word that means Moon because there is one Moon.

A common theme throughout the Vedas is the connection between the stars, life and the spirit. Vedic ritual was based on the times for the full and new Moons, solstices and equinoxes. The oldest existing astronomical text from India is the *'Vedanga Jyotisha'* from the end of the Vedic period, spanning 1500BC – 500BC. The *'Surya Siddhanta'* is another important ancient text dating from the 6th century BC.

In the Vedic eclipse myth I mentioned in the first chapter, Rahu and Ketu are considered as two dark planets that are responsible for the eclipses. Astronomically, they do not exist

---

[62] You can find a good summary of non-western astronomy in the 2008 book 'Astronomy across cultures: the history of non-western astronomy' by Helaine Selin.

but are shadow entities – the two points of intersection of the lunar path with the ecliptic. These are the places where the Moon needs to pass in order for an eclipse to occur. In the Vedic astrology of the Jyotish-Shastra, the Sun and Moon were believed to have a strong impact on our lives, the Sun representing the body and the Moon representing the mind. As Rahu and Ketu cause the eclipses, they were thought to effect the energies of body and mind.

Vedic astronomers tracked the motion of the Moon against the backdrop of the constellations they called 'nakshatras' to set their lunar sidereal calendar. The nakshatra system dates to before 1000BC, and a set of 27 stars were used as reference points for noting the position of the Moon as it moves through its 27.3 day lunar sidereal cycle. During any individual day the sky was also divided into 27 parts, each 13.33 degrees wide. The ancient texts describe the lunar-solar calendar, length of the year, dividing circles into 360 degrees and the ratio of longest to shortest day being 3:2. Eclipses were also mentioned, as well as the method of calculating time using a gnomon.

In the Rigveda the universe is described as infinite and very old, its creation occurring 4.32 billion years ago.[63] The sequence 432 often appears in Vedic mythology. The Rigveda itself is said to consist of 432,000 syllables. These texts do not mention astrology or prophecy but are focused on timekeeping, particularly for tracking the Vedic rituals.

---

[63] "Scriptures, science and mythology: Astronomy in Indian Cultures", 2009, R. Kochbar, Proceedings of the International Astronomical Union, 260.

Soma, or the Moon, is one of the most important deities of the Rigveda. An ancient myth explains why the Moon wanes. According to the Brāhmaṇa texts, the creator god Prajāpati gave his daughters, the nakshatras, in marriage to the Moon. However, the Moon neglected all his other astral wives and cohabited only with Rohiṇī, his favourite wife. The other wives in anger returned to their father, who severely reprimanded his son in-law. The Moon promised to treat all his wives equally, but still continued cohabiting with just Rohiṇī; in punishment he was inflicted with the illness that makes him wane.

### Astronomy in ancient China

Recent archaeological discoveries have confirmed the importance of astronomy in the early history of Chinese civilisation. Carvings of the Sun, Moon and stars dating from 6000BC were discovered at Damaidi in China. In 1988, a tomb dating to 4000 BC was excavated in Puyang County, Henan province. Its occupant was surrounded by a group of figures made up of clam shells – a dragon to the east, a tiger to the west, and a ladle to the north. Archaeologists believe that this configuration is a representation of the Chinese sky: the ladle represents Ursa Major, a major constellation in the circumpolar region; the dragon is the eastern asterism Blue Dragon; and the tiger the western asterism White Tiger.

Some of the earliest writings of ancient China on astronomy were inscriptions carved onto bones, known as 'oracle bones' for their use in fortune telling. The practice included writing an observation or question on the bone and placing it in a fire. The cracks in the bones were then used to foretell the future. Astronomical observations include mentions of 'guest stars', which were supernovae, the spectacularly bright deaths of stars. Star names later categorised in the twenty-eight mansions have been found on oracle bones unearthed at Anyang, dating back to the middle Shang Dynasty around 1200BC.

The '*Canon of Yao*' originates in China around 2400BC and discusses how the 6pm observations of the meridian passage of groups of stars were used to record the times of the equinoxes and solstices. The use of a precise time of day means that fairly accurate time-keeping devices had to be used, such as a candle or water clock. Although it is difficult to interpret, it mentions that the brothers His and Ho were commissioned by the legendary emperor Yao to define the calendar: *"to conform themselves to august heaven, to calculate and plot the Sun, the Moon, the stars and the celestial bodies and respectfully to submit a calendar for humankind."*[64]

Just as in Babylon, in ancient China it was believed that events in the sky directly reflected events on Earth. Comets were particularly important omens. Halley's Comet appears in the sky every 70 to 75 years, so very few people see it more than once in their lifetime. Over time, Chinese astronomers recorded every instance of Halley's Comet from 3000 years ago, the only civilisation in the world to have done so. They also made drawings of how the comet looked each time it visited the vicinity of the Earth and Sun.

---

[64] "Connecting Heaven and Man: The role of astronomy in ancient Chinese society and culture", 2009, X. Sun, Proceedings of the International Astronomical Union, 260.

Eclipses were often used to determine the future health and welfare of current emperors and empires. The emperor's astronomers were responsible for producing the calendar each year, a document commonly known as an almanac. No one else was allowed to calculate a calendar. Royal astronomers were in a difficult position. If they failed to predict an event like an eclipse then the emperor's power might appear diminished, and political rivals could take it as an opportunity to rebel. When dynasties fell, it was believed that Heaven had directly intervened to give the responsibility of rule to a more worthy line. The job of royal astronomer therefore carried huge responsibilities, and severe penalties were given for any mistakes.

The superstitious belief that linked events on Earth to those in the heavens made people very wary of phenomenon in the sky. Solar eclipses in particular were regarded with fear, and a common belief was that these occurred because a great dragon was attempting to devour the Sun. The forewarning of such an event was therefore imperative so people could gather to shout, strike gongs and scare away the dragon. Over generations of observation, Chinese astronomers also discovered the Saros cycle. This enabled them to predict solar and lunar eclipses with some accuracy but, as we will learn in the next chapter, it was not an infallible system. In 2136BC there was an unpredicted total eclipse of the Sun. Documentation about this event dates it as the earliest recorded eclipse in history, but it also tells us about the fate of the court astronomers Xi and He who failed to predict it in advance. Given the belief that such celestial events reflected events on Earth and should be predicted by the emperor, complete accuracy was expected of court astronomers, and failure meant only one fate – execution. I should tell my PhD students this story!

The belief which linked celestial activity to that on Earth is illustrated in a description of lunar behaviour by a court astronomer, Shishen, in the 4th century BC: *"When a wise prince occupies the throne, the moon follows the right way. When the prince is not wise and the ministers exercise power, the moon loses its way. When the high officials let their interests prevail over public interest, the moon goes astray toward north or south. When the moon is rash, it is because the prince is slow in punishing; when the moon is slow, it is because the prince is rash in punishing."*[65]

The Sun and Moon have always had a special significance in Chinese folklore, and various symbolisms and myths surround them. The mid-Autumn, or Moon Festival is perhaps the second most important traditional festival in China after the Chinese New Year. It is held on the fifteenth day of the eighth lunar month when the Moon is said to be at its largest, roundest, and brightest of the year. The shape of the Moon is thought to represent completeness and perfection, and its celebration is an important family occasion. Special round cakes, called Moon cakes, are made to eat during the festival.

In Chinese folklore, the features on the Moon are associated with a white hare or rabbit, sometimes pounding a pestle and mortar. One Buddhist story tells that Buddha was once a hare and sacrificed himself to the god Indra who was

---

[65] H. Maspero, 'L'astronomie chinoise avant les Han' ['Chinese Astronomy before the Han Dynasty'], T'oung Pao 26 (1929): 288

suffering from hunger. As thanks for his selflessness, Indra then immortalised the hare by placing its image on the Moon for all to see. The first successful lunar rover that the Chinese landed on the Moon in 2013 was named Yutu, the 'Jade Rabbit'.

It is interesting that early Chinese astronomy and the search for omens most likely developed independently from Babylon. In the MUL.APIN, the Babylonians divided the path of the Moon into 17 or 19 stations, which by the 5th century BC had evolved into the 12 sign zodiac that is still used today in the West. The Chinese divided the path of the Moon into 28 lunar mansions, and the remainder of the stars in the night's sky into over one hundred asterisms. Trade between China and the Indian subcontinent may be reflected in their similarity of constellations. In the Vedanga Jyotisha dating from the final centuries BC, groups of stars along the ecliptic were divided into 27 or 28 lunar nakshastras.

**Astronomy in isolation**

The similarity of the Babylonian, Chinese and Indian constellations of the zodiac, which lie on the ecliptic, suggests that some knowledge may have been passed between these early civilisations. Long distance trade was certainly occurring by the 3rd millennium BC. The gemstone lapis lazuli was being transported from its only known source in the ancient world in north-eastern Afghanistan as far as Mesopotamia, Egypt and the Indus Valley. However, the astronomy of the Mesoamerican cultures was truly independent from Europe and Asia. It is therefore interesting to see if calendar systems, eclipse omens and astrology developed independently.

Asian nomads entered the Americas at least 15,000 years ago and possibly much earlier. After around 2000BC complex cultures began to form in South America, including the Olmec, Maya, Zapotec, and Aztecs. These indigenous civilisations constructed pyramid-temples and developed mathematics, astronomy, medicine and writing. They also used highly accurate calendars, created fine arts and developed intensive agriculture, engineering and abacus calculators.

But all was not so great. Many of the South American cultures held some rather extreme beliefs. Unlike the Incas, the Chimú culture in Northern Peru worshipped the Moon, believing it to be far more powerful than the Sun. But like the Incas they also sacrificed animals and children to the gods. In 2018 archaeologists finished their excavations of the Huanchaquito-Las Llamas temple and counted the remains of 200 young lamas and 140 children that had been sacrificed in one event. Why, or to which deity they were hoping to appease, is not known.

Little is known about the history of these civilisations since almost all of their texts were destroyed by the conquistadors and Catholic priests in the 16th century who did not find their religion appealing. Their aim was to convert them to Catholics, and together with the expansion of the Spanish empire they ended up causing one of the largest genocides in human history, resulting in the death of millions of natives. Only four ancient texts exist, called the 'Maya codices', folding books written on cloth made from the bark of trees that give us a tiny glimpse into their early culture. It is like trying to decipher the history of the ancient Greeks with a random collection of four short books. Three of the four codices are primarily about astrology, but they also contain information about eclipses, Venus cycles, the Mayan zodiac and prophecies based on the Mayan calendar.

The Mayan calendar is particularly complex, based on a 260 day ritual calendar and a 365 day yearly calendar. Their months were 20 days long, perhaps because of the base 20 number system they adopted. Or perhaps this was because the Mayans divided the sky into twenty constellations. It is not known if the 365 day yearly calendar was corrected using intercalary days. If not, after a few centuries it would have been completely out of synchronisation with the seasons.

These two calendars were used together with a 52 year 'short count' calendar round, which is the period of time for the two former calendars to synchronise. Their calendars had a starting point of a day in 3114BC, the day the world was thought to have been created. In addition, there is a long count calendar, which was used for longer time periods. Misinterpretation of the Mayan calendar was the basis for a popular belief that the world was going to end on December 21, 2012. However, that date was just the end of another long count period.

It is interesting that in this isolated part of the world, a completely different number system and calendar arose. This calendar was also used by the Aztecs and other Pre-Columbian civilisations across Mesoamerica and probably began with the Olmecs before 500BC. It is not known why a 260 day calendar for rituals was used, nor the significance of their long count calendar, but of course there are many theories. Unless further texts surface we may never know more about the astronomical methods of these fascinating civilisations.

Together with inscriptions on walls and pottery, some of the knowledge of the Mayans has been reconstructed. They carried out accurate measurements of the lunar and solar cycles, eclipses and motions of the planets. Their estimate of the synodic lunar month was even more accurate than the Babylonians'. They measured the orbital time of Venus as 583.92 days, accurate to a precision of 14 minutes. The tropical year was measured by the Maya as 365.242 days. Compared to the modern value of 365.24198 it is correct to within 18 seconds! Just like with the Babylonians and the Indian and Chinese cultures, the huge effort in making these accurate astronomical measurements did not seem to be matched by any attempt to understand the workings of the cosmos. Instead, the heavens were the home of the deities and treated as an immense clockwork mechanism to predict the future.

Before describing the attempts of the first philosophers to make sense of the world, let's look at the workings of our clockwork-like solar system and see just how close these ancient civilisations were in their ability to predict the eclipses of the Sun and Moon.

# 7. Illusions of Light and Shadow

If there is one celestial phenomenon that captures our imaginations more than any other it is the eclipses of the Sun and Moon. One can only imagine that our distant ancestors must have thought the world was going to end when the daytime sky turned into night as the Moon passed in front of the Sun. It must have also been more than a little worrying when the Moon slowly faded and turned a dark blood red for an hour or more as it passed through the shadow of the Earth. It is not surprising that these ominous events were associated with omens and forthcoming disaster. After all, in many mythologies, the Sun and the Moon were associated with the gods and creators of the universe, so the eclipses had to be some sort of a message.

The reason that the Moon turns red during a lunar eclipse was not solved until the 17th century, when Johannes Kepler explained that it was the refraction of sunlight through Earth's atmosphere. The blue wavelengths are scattered in all directions whilst the red light passes through the atmosphere and is refracted towards the Moon. It's the same reason that the sky glows red even after the Sun has set on the horizon. If Earth had no atmosphere, the Moon would turn completely black during such an eclipse, and it would be invisible against the night's sky.

### Timely knowledge

The Babylonians could accurately predict when an eclipse of the Moon would take place and when it could be seen. They could also predict when an eclipse of the Sun would occur, but not where it would take place. It took over two thousand years before astronomers learned how to calculate where a total eclipse of the Sun would occur. That might seem like slow progress, but quite a lot of science had to be figured out first. The type of eclipse and the time between eclipses depend on the orbit of the Moon. But the path of our Moon is extraordinarily complex. So much so that understanding the motion of our Moon puzzled the greatest scientists from Hipparchus to Isaac Newton.

Writing in the 4th century BC, the Greek historian Herodotus recounts a historical battle in Asia Minor between the Medes and the Lydians that came to a halt when a total eclipse darkened the sky. Following the event, the two nations made peace, believing the eclipse was a sign for them to stop the fighting. Herodotus also mentions that the eclipse was predicted in the same year by the Greek philosopher Thales.

However, the ability to predict the time and location of an eclipse of the Sun would have been remarkable at this time in history. It would require very accurate knowledge of the motions of the Sun and Moon, which did not exist. Even if Thales had heard of the famous Saros cycle of eclipses that the Babylonians discovered, that would be insufficient knowledge to predict a total eclipse of the Sun. Herodotus was describing events that took place a century before. Astronomers have calculated that a total eclipse of the Sun did occur in the year 585BC at the time of Thales. However, it may have been that the battle was being fought by the light of the full Moon and it was the Moon that turned dark during a lunar eclipse. Eclipses of the Moon could be predicted using the knowledge of the Babylonians. That makes much more sense to me, especially since Herodotus recounts other battles between the Medes and Lydians taking place during a full Moon.

The story of Christopher Columbus using the knowledge of a forthcoming lunar eclipse to impress the natives on the island of Jamaica is widely known. Lunar eclipses could be predicted well in advance at that time, and Columbus had an almanac containing astronomical tables covering the years 1475-1506. It was made by Johannes Müller von Königsberg, a highly regarded German mathematician and astronomer. The almanac provided detailed information about the Sun, Moon and planets, as well as the more important stars and constellations to navigate by. Any respectable sailor at this time would have a copy of Müller's almanac. Columbus noted that on the evening of Thursday, February 29, 1504, a total lunar eclipse would occur, beginning around the time of moonrise.

He warned the reluctant locals that if they did not provide his ship with supplies, his christian god would be angry and show displeasure by turning the Moon red with rage. As expected, the lunar eclipse occurred and the Moon turned the colour of blood red. According to Columbus' son, Ferdinand, the Arawaks were terrified at this sight and *"with great howling and lamentation came running from every direction to the ships laden with provisions and beseeching the admiral to intercede with his god on their behalf."* They promised that they would happily cooperate with Columbus and his men if only he would restore the Moon back to its normal self.[66]

In 1715, the astronomer Edmund Halley published a map predicting the time and path of a coming solar eclipse over Southern England. The time was correct to four minutes and the path was accurate to 20 kilometres. Halley's was not the first eclipse map but it was the first one that was very accurate and widely sold to the public. The earliest confirmed solar eclipse map was calculated by Erhard Weigel, an important figure in the German enlightenment. It showed the path of a total eclipse that crossed Europe in 1654 and was published the day before the eclipse.

Even at this time, total eclipses were still rather scary events. Hans Heinrich Bluntschli in *'Memorabilia tigurina'* (1742) describes such an event that took place across Switzerland and that he observed in Zurich: *"The terrifying solar eclipse which occurred on 12 May 1706 is still fresh in our memory. It began at eight forty-five in the morning; was in the middle, that is at its biggest, at ten o'clock, and ended at eleven o'clock, in which time the Sun had been masked by the moon as if with a curtain, and thus lost its glow to such an extent that the stars even came out. By around ten o'clock, no more work could be done. The chickens retreated into the tranquillity of their nests and bats could be seen flying around Weinplatz.* Bluntschli ends his story with a brand new eclipse omen: *"This murkiness was followed by a very hot summer, and all was well with the wine and fruit."*[67]

---

[66] "The life of the admiral Christopher Columbus by his son Ferdinand", 1959, translated by Benjamin Keen, Rutgers University Press.
[67] "Memorabilia tigurina", 1742m H. H. Bluntschli, translation ETHZ library, Zurich.

### Predicting the eclipses

Now it's time for me to explain in a little more detail at the phenomenon of eclipses.

Eclipses can only happen when the Moon, Earth and Sun are arranged in an exact line, such that the Moon obscures the Sun or the Moon passes through the shadow of the Earth. This alignment has the wonderful name 'syzygy', which is great to remember during games of scrabble. At these times the Moon is crossing the ecliptic and is either in its full or new phases. If the Moon orbited the Earth exactly in the ecliptic plane, then there would be a solar and lunar eclipse every month. But the Moon orbits the Earth inclined to the ecliptic, and that orbit is an ellipse which is constantly changing its orientation.

When the Moon is closest to Earth it is at a point called perigee (363,000km), when it is furthest away it is at apogee (406,000km). At perigee the Moon is closer to Earth and appears ten percent larger than at apogee – a so called 'supermoon'. The points of perigee and apogee are at the ends of the elliptical orbit. It is a coincidence that the angular size of the Moon is about equal to the angular size of the Sun and can therefore block out all of its light during a total eclipse. But the type of solar eclipse that we see, whether it is partial, total or annular, depends on the distance from the Earth to the Moon during the eclipse, and this varies because of the Moon's elliptical orbit.

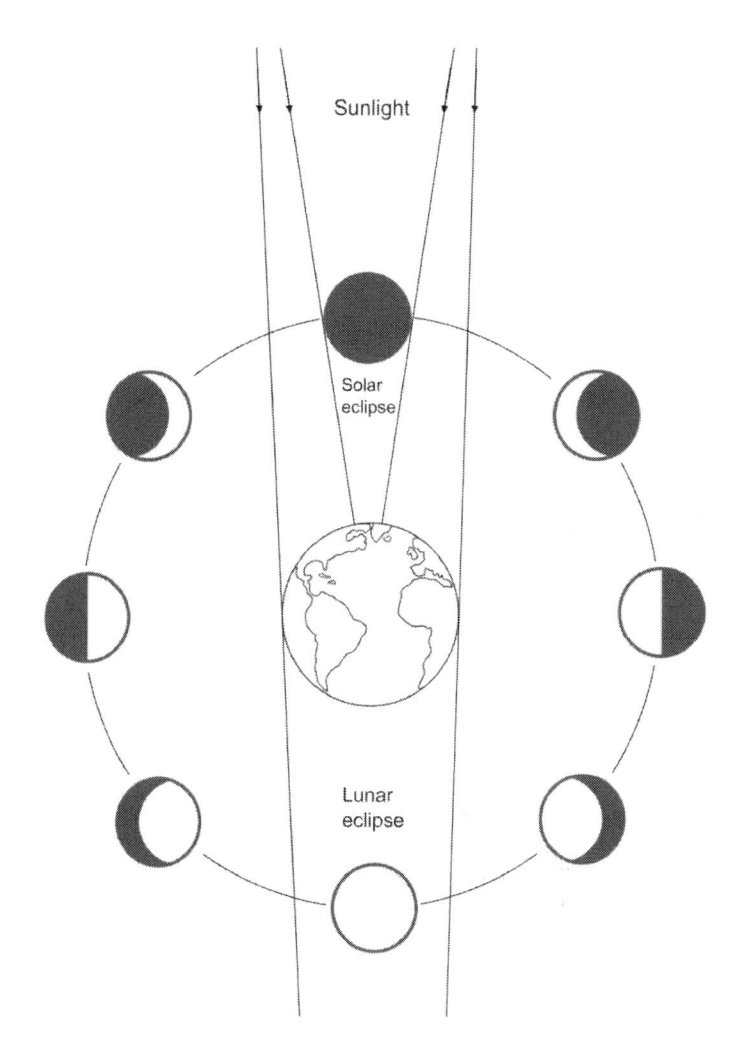

*Solar and lunar eclipses*

On average, there is a total eclipse of the Sun and Moon somewhere on Earth about once every 18 months. The shadow of the Earth at the distance of the Moon is several times the size of the Moon. Therefore, a lunar eclipse lasts between one and two hours and is visible from a large region of the world at night. However, during a solar eclipse, the Moon's shadow on the Earth is only about 100 kilometres across and the Moon is moving around the Earth at about one kilometre per second. Therefore the Moon's shadow is moving across the surface of the Earth at about 2000 kilometres per hour, and a total solar eclipse lasts only a few minutes. And because of the complex path of the Moon and the small size of the shadow, a solar eclipse will typically only occur at the same location on average after about 360 years.

If an eclipse has occurred at a certain time and place, there are three important timescales that need to synchronise in order for the alignment to reoccur. We have already come across the first one. The time between the same phases of the Moon is the *synodic month*, 29.53 days on average, but this can vary by several hours because its orbit is not circular. This timescale would have been known to any careful Neolithic observer of the night's sky.

To complicate things further, the Moon's elliptical orbit precesses around the Earth once every 8.85 years. This means that over this time the perigee and apogee points of the Moon's elliptical orbit slowly rotate around in a complete circle. This is called the apsidal precession and is due to gravitational perturbations from the Sun. Because of this slow rotation of the elliptical orbit, the time it takes for the Moon to travel from perigee and back to perigee is slightly longer than its orbital time around the Earth. This period of 27.55 days is called the *anomalistic month*. The 2$^{nd}$ century BC Greek astronomer Hipparchos noted the 8.85 year apsidal precession of the Moon and it is corrected for in the famous 'Antikythera Mechanism'.

The final timescale that is needed in order to understand when eclipses occur is the *draconic month*, sometimes called the nodal month. The Moon's orbit about the Earth is tilted to the plane of the ecliptic (the plane defined by Earth's orbit about the Sun) by an average of about five degrees. Eclipses can only occur when the Moon crosses the plane of the ecliptic, because then it is in the plane of Earth's orbit and can pass in front of the Sun or through Earth's shadow. But the plane of the Moon's orbit also precesses, i.e. it tilts up and down by about ten degrees over a timescale of 18.6 years. This means that the time taken for subsequent ecliptic crossings is shorter than its orbital period (the sidereal month), and is the draconic month of 27.21 days. This is called the nodal precession and is due to the fact that the Earth is not exactly spherical, which results in an additional varying gravitational pull on the Moon. Now you can understand how difficult it was to make sense of the motion of the Moon!

(i) Top down view of the lunar orbit

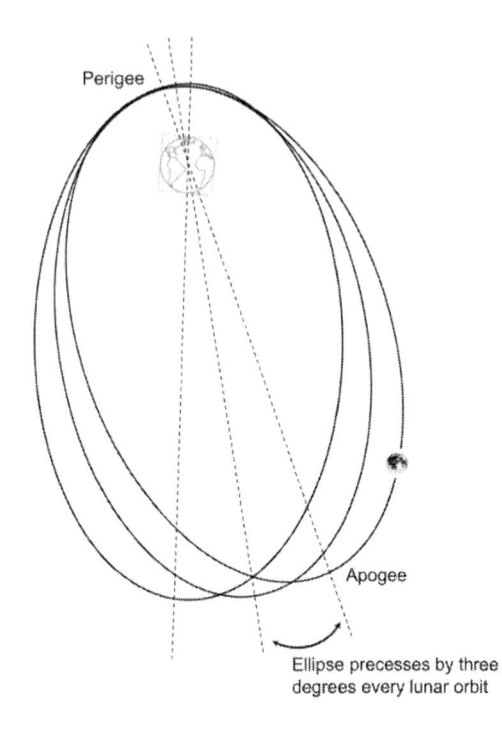

Perigee

Apogee

Ellipse precesses by three degrees every lunar orbit

(ii) Edge on view of the lunar orbit

The Moon's orbit tilts up and down ten degrees over a timescale of 18.6 years.

*The complex lunar orbit, viewed from the top and the side*

The Babylonians accurately measured these different monthly timescales as well as the mean lunar speed with respect to the stars and its monthly accelerations and decelerations. To give you an idea of the incredible accuracy of the Babylonian astronomers, their estimate of the synodic month was 29.530594 days, which is only four seconds different from the modern value! Their value of the synodic month is still in use as the basis of the Hebrew calendar today. The precession of the Moon's orbit in inclination and angle were also known from ancient times, but were not characterised as such until Kepler and Newton.

Eclipses will reoccur when the cycles of the draconic month, the synodic month and the anomalistic months coincide. So we need to search for multiples of these periods that are identical or at least very close to the same number of days. The first eclipse cycle happens because 223 synodic months is 6585.32 days, 242 draconic months is 6585.46 days and 239 anomalistic months is 6585.45 days. This period is 18 years 11 days and 8 hours, and is known as the Saros cycle. So given that an eclipse has happened on a certain date you can guess that another one will happen a Saros cycle later. And because the cycle length is just over 11 days longer than a year, the tilt of the Earth will be close to its same position again, so the eclipse will occur very close to same latitude.

Lunar eclipses can be seen from most of the side of the Earth that faces the Moon, therefore knowledge of the Saros cycle is all that was needed to predict when the next lunar eclipse will take place. However, because the Saros period is not an exact integer number of days – it has an extra eight hours – successive solar eclipses don't occur in the same part of the world (at the same longitude). That's because the Earth has rotated by an additional 120 degrees in a Saros cycle. This gives rise to an even longer cycle, the triple Saros cycle, in which eclipses occur near the same part of the world (latitude and longitude) every 19,756 days. Because none of the timescales match Earth's rotational speed exactly, the precise location on Earth of a solar eclipse varies. But for a kingdom such as Mesopotamia, spanning a region over a thousand kilometres, any one Babylonian king may have had reports of several solar eclipses during their lifetime.

After the careful refinement of the celestial spheres model for the lunar and planetary motions, it became possible to make rough predictions for the place and time of solar eclipses. The 'Zij' were Islamic astronomical books produced between the 8th and 15th centuries based on Ptolemy's model for predicting the positions of the celestial bodies. They could accurately predict the lunar eclipses, but the solar eclipses were still hit and miss. The causes and effects of the complex lunar motion were not fully worked out until the 20th century. However, the basic ideas fell into place in the 16th century with the observations of the Moon's motion by Tycho Brahe and the empirical theory of orbits by Johannes Kepler. Only with this knowledge could astronomers start to make accurate predictions of when and where a solar eclipse might occur.

**Shades of grey**

It must be a beautiful sight to stand on the Moon and witness an eclipse of the Sun by the Earth. None of the Apollo missions coincided with such an event, but a robot witnessed this in 1967, when NASA's Surveyor III lunar lander captured one photograph of the Sun disappearing behind the Earth, the last remaining sunlight passing through the tops of Earth's high-altitude clouds.

The sight of the Earth from the Moon was described as spectacular and humbling by all the Apollo astronauts. Alan Shepard recounts standing on the Moon during the Apollo 14 mission in 1971: *"When I first looked back at the Earth, standing on the Moon, I cried."* Because of Earth's size and the reflectivity of its oceans, clouds and ice-caps, a 'full Earth' as seen from the Moon is four times as large, and is over 40 times brighter than the full Moon as seen from Earth! The Earth viewed from the Moon is a vibrant blue and white globe framed by the blackness of space.

We are all familiar with the view of our Moon from Earth, but what colour is the Moon? To our eyes the Moon appears a bright silvery white colour. But if you look at the photographs taken from the surface of the Moon, you will see that the surface looks a rather dark grey. Apollo astronaut Charles Duke described his experience of standing on the Moon: *"My most vivid memory on the moon is the beauty: the stark contrast between the brilliant gray of the moon and the blackness of space. The gray was so bright it was almost white – a sharp break between the surface and the horizon. The Sun was always shining, so you didn't see stars or planets."*[68]

The way our brain interprets signals from our eyes is remarkable and responsible for a host of confusing illusions.

---

[68] "What's It Like To Stand On The Lunar Surface?" 2017, J. Clash, Forbes.

The Moon colour paradox is another one of them. Grey is not a colour but a shade of white. We perceive grey when the cones in our retina receive equal amounts of red, green and blue light. The shade of grey we perceive also depends on the brightness of the surroundings. This is nicely illustrated by the contrast illusion in which two identical grey squares are placed within a different background. The grey square that lies in a shaded region of the image appears to be a much lighter shade of grey. Likewise, the grey Moon appears a silvery white because we view it against the blackness of the night's sky.

Not convinced? I wasn't. So I thought about a way of demonstrating this and tried the following experiment. The full Moon has an angular size of about half a degree. Cut a circular hole that is about one centimetre in diameter from a sheet of white paper. When you hold this a metre away, the hole will have an angular size the same as the Moon. At night, when the Moon is full, hold the paper at arm's length, such that the Moon is visible through the hole. The Moon will look its usual bright silvery white colour. Now shine a bright flashlight onto the paper, bright enough that the white paper is illuminated as bright as daylight. You will see that the Moon looks grey in comparison to the white paper! Turn off the light, and within a second the Moon appears bright white again.

What is the mechanism behind this strange illusion – is it our brain or our eye that is deceiving us? Our highly complex eyes actually do a lot of image processing even before the signal reaches our brain. The response of each point on the retina is influenced by neighbouring regions. The signal that reaches a ganglion cell on the way to the brain contains information from the rods and cones in a certain area of the retina. When light falls onto a photoreceptor in the eye, it responds by firing signals towards the ganglion cells more frequently, and it also inhibits adjacent cells from firing. This is called lateral inhibition, and it is a mechanism that enables us to emphasise contrast – presumably an evolutionary survival advantage. When the object is uniformly illuminated this decreases the overall firing so that all of the white is not as bright. But when there is a transition from dark to bright, the receptors next to the dark edge are suppressed. So grey surrounded by white appears a darker shade of grey than grey surrounded by black.

Our eyes receive the same information from a dimly illuminated object that perfectly reflects light as from a brightly illuminated object that only reflects a fraction of the light. So what colour is the Moon? The Moon shines from reflected sunlight, so we first have to determine the colour of the Sun. The broad spectrum of wavelengths that the Sun emits peaks in the green frequencies. But sunlight contains a wide spread of colours, an almost equal mix of red, green and blue light, and that's why we perceive the Sun as white. And that is why the Moon appears white, or a grey shade of white.

But what is the 'whiteness' of the Moon? How white something is, is determined by the amount of light that object reflects and that is called its albedo. A perfect mirror would have an albedo of one, whereas black paint typically reflects just five percent of light and its albedo is 0.05. The albedo of a planet tells us a great deal as to its atmosphere or surface.

Forests have an albedo of 0.1-0.2, sandstone 0.2, ocean ice 0.6 and fresh snow 0.8-0.9.

The albedo of the Moon was first accurately measured by the German astronomer Karl Friedrich Zoellner in the mid-19th century. It is difficult to infer because the Moon is a curved surface covered with irregularities that scatter the light or create shadows. So a crescent Moon, in which say ten percent of the edge of the Moon's surface is illuminated, is far less than a tenth of the brightness of the full Moon.

The first attempts to measure the brightness of the Moon used observations of the Sun and Moon projected to a point and comparing the intensity of that image to the light from candles. Zoellner found that the light of the full Moon was about 600,000 times fainter than the Sun. After considering the known size of the Moon and its distance from the Sun, he estimated that if the Moon were a perfect white disk it would be about 100,000 times fainter than the Sun. He therefore estimated that the Moon reflected just one sixth of the Sun's light and remarked that the Moon should be regarded as closer to black than white.[69]

Prior to Zoellner's estimate of the Moon's albedo, the British astronomer Sir John Herschel had deduced the albedo of the Moon in a much simpler way. Herschel had looked at the full Moon setting behind Table Mountain (the location of his observatory when he went to observe the stars in the southern hemisphere) and noted that it was indistinguishable

---

[69] "The Moon: her motions, aspect, scenery and physical condition", Richard Anthony Proctor, 1873, Longmans, Green and Co. London, page 237.

from the grey rock it moved behind. Since the rocks of Table Mountain are weathered sandstone with an albedo of less than 0.2, Herschel's by eye estimate of the Moon's albedo was not far off.

The surface of the Moon is actually a poor reflector of sunlight – its albedo is 0.12, which means that only around 12 percent of the incident light is reflected. That's why the Moon's surface is a dark shade of grey. The albedo of the Earth is about 0.3, which was first measured by observing the brightness of Earthshine, the light reflected from the Earth to the Moon and back, long before a camera had been sent into space!

The albedo of the Moon is lower than the Earth's because there are no oceans or large patches of ice on the Moon. But the reflectivity varies across the surface of the Moon due to the geology of the rocks on the surface. The dark parts are the maria, basalt rock which formed long ago as lava flowed onto the surface and cooled. The lighter parts are the terrae, or highlands, made primarily of anorthosite and breccia which reflect more light and appear a lighter shade of grey.

**The illusion of the rising Moon**

It is a beautiful sight to witness the full Moon rising over the horizon. It appears to be far larger than when it is high in the night's sky. Most people estimate that the Moon appears to be almost twice as large when it is on the horizon as opposed to overhead. This was a great puzzle in ancient times, and you may be surprised to learn that scientists and philosophers are still puzzled about it today.

The so called 'Moon illusion' has been known about for thousands of years. It was mentioned by the early Chinese scholar Lieh-Tseu in the 5th century BC and by Aristotle in the *Meteorologica* in the 4th century BC, who thought it was an effect of the atmosphere. Ptolemy took this further in the *Almagest*, arguing that when the Moon is low in the sky, the reflected light passes through more of Earth's atmosphere, which refracts the light as a lens making the Moon appear larger.[70] These explanations are not correct. In fact, because of the viewing geometry, the Moon has an apparent size that is slightly larger when it is high in the sky rather than on the horizon. That's because it is slightly closer when it is overhead, but only by about 1.5 percent, and that is not noticeable by eye.

In the 11th century, the Arab mathematician Ibn al-Haytham argued that the Moon illusion was an effect of perceived distance. This explanation has some variants. The most common notion is that when the Moon is close to the horizon it can be seen next to objects like trees or houses that give our brains something with which we can compare the size of the Moon. This explanation was endorsed by many early scientists and philosophers, such as Roger Bacon, René Descartes and Johannes Kepler. However, in 1762 the Swiss mathematician Leonhard Euler made the point that the illusion persists even when the rising full Moon is seen over a featureless ocean or desert.

In 1678 the theologian philosopher Nicolas Malebranche argued that the Moon illusion demonstrated the unreliability

---

[70] "Ptolemy's theorem on the apparent enlargement of the Sun and Moon near the horizon", 1900, T.J.J. See, Popular Astronomy, 8, 362.

of sensory information in general. He went on to state that that man should distrust his senses and his reasoning, and rely upon the authority of the church for truth!

The most common scientific explanation today is that the horizon Moon illusion is due to how we perceive distance as proposed by Ibn al-Haytham. But not by the comparison with nearby objects, but by our brain's learned model of the space around us. For nearby objects we can use the parallax of our dual eyes to determine distances. Our eyes are around six centimetres apart, which gives us two slightly different views of the world. Our brains learn to measure distance by the change in position of the nearby object compared to distant objects. That's called parallax. Hold your finger up and look at how the background changes when you use different eyes. But the angular resolution of our eye is about 1/60th of a degree, so this technique is only useful for objects less than 30 metres away.

We perceive the distances to more distant objects by experience – by comparing perspective and constructing a model of the world in our brain. Our local environment on Earth is flat, and we use perspective to infer distance. We think that the Moon looks larger when on the horizon because we believe that objects on the horizon are further away. And that is generally the case for most of the things we look at, but not in the case of the Moon. Take clouds as an example. Above our heads, clouds are typically a few kilometres away, but the same clouds on the horizon can be seen to far larger distances. We therefore build a perception of the sky as being closer to a flat surface overhead rather than a spherical bowl. Therefore, our brain tells us that objects on the horizon should appear smaller. If we saw a cloud on the horizon that had the same apparent size as a cloud overhead, we would correctly think the cloud on the horizon was enormous. And that is thought to be the solution to the Moon size paradox – when the Moon is on the horizon it has the same size as overhead, but our brains are telling us that it is far away so it must be larger.

This explanation has been discussed extensively in the scientific and philosophy research journals. Experiments have been carried out using fake Moons against different backgrounds, at different angles in a planetarium, looking with one eye, or viewing the rising full Moon through a tube so the landscape is not visible. The physiological details of why we perceive the Moon to be larger are still not agreed upon, and there are many proposed explanations.[71,72] Some are psychological, others are mechanical in the sense of how our eye responds to light and angle. So, despite all the experiments and advancements in science, we still do not have a consensus on the true nature of how the Moon illusion manifests itself in our brain!

---

[71] "The Moon illusion examined from a new point of view", 1975, J.T.Enright, Proceedings of the American Philosophical Society, 119, 2, 87.
[72] "The Moon illusion: Kaufman and Rock's (1962) apparent distance theory reconsidered", 2007, K. Suzuki, Japanese Psychological Research, 49, 1, 57.

## 8. Mediterranean Musings

Prior to the 6th century BC our Moon was a shining light, a mystical object of worship and bearer of omens. The Earth was a flat disk, and the Moon and Sun chased each other across the sky driven by chariots of the Gods. Would you have come up with a different view of the world if you were not given knowledge and explanations by others? Quite suddenly, two and half thousand years ago on the borders of the Mediterranean, our understanding of the world and the cosmos changed thanks to some incredible philosophers. Some of their achievements were enabled by the precise observations and systematic record keeping from the Babylonians. Other insights were made by thought alone.

Thus began the period when the ancient Greeks tried to make sense of it all. Some failed spectacularly and set science back by a millennia, others made spectacular insights that set us on the path of discovery towards understanding the cosmos. However, the single greatest achievement of the first Greek philosophers was breaking away from explaining natural phenomenon as something due to the gods. Events in the skies were not omens but natural phenomenon that could be explained by logic and science, by cause and effect. This period begins with Thales of Miletus in the 6th century BC and ends with Hipparchus of Nicaea in the 2nd century BC, at the beginning of the expansion of the Roman Empire.[73]

---

[73] "A history of astronomy from Thales to Kepler", 1953, J.L.E.Dreyer, Dover publications Inc, London.

In Mesopotamia, war after war between neighbouring tribes took place until in 539BC Babylonia was absorbed into the Persian Empire. During the 26th Dynasty of Egypt (c. 685–525 BC), the ports of the Nile were opened to Greek trade for the first time, and the first ancient Greek philosophers such as Thales and Pythagoras visited Egypt bringing with them new ideas and taking Egyptian and Babylonian knowledge back to Greece. Power shifted from Asia to Athens, whilst at the same time the Persian Empire was taking over the Assyrians' and began its wars with Egypt. This is the so called classical Greek period, which lasted until the death of Alexander the Great in 323BC.

Very little is known about ancient Greek, 'pre-Socratic' philosophy. Nearly all original works have been lost, with only fragments of knowledge surviving, mainly thanks to references by later authors such as Aristotle, Herodotus and Ptolemy. The 6th century BC philosophers such as Thales, Pythagoras, Parmenides, Anaxagoras, Anaximander and Anaximenes, to name a few, were amongst the first to question the nature of the cosmos. They identified the cause of eclipses and the fact that the Moon reflected a portion of the Sun's light and the Earth was a sphere. But some of their views made little more scientific sense than some of the Babylonian cosmological stories that were incorporated into the Old Testament.

Aristotle identified Thales as the founder of natural philosophy and the first person to question the nature of the world. Thales believed in a materialistic world in which there was cause and effect. In the 6th century BC Thales questioned the basic material of the cosmos, a question that myself and many of my colleagues still try to answer today, albeit still with little success. Plato had Socrates relate a story that Thales was so intent upon watching the stars that he failed to look where he was walking and fell into a well!

## The nature of the Moon

In the late 19th century, hundreds of boxes of decaying papyrus documents were found in an ancient Egyptian rubbish dump. Named the 'Oxyrhynchus Papyri', and dating from the 3rd century BC to the 6th century AD, the texts are mainly written in Greek. Only a tiny fraction of the documents have been transcribed due to their condition. They cover a range of subjects, from tax assessments to private letters. One fragment that was translated and published in 1986 reveals the first person to understand the mechanism of eclipses and speculate as to the nature of the Moon: Aëtius testified that Thales *"says that eclipses of the Sun take place when the Moon passes across it in a direct line, since the Moon is earthy in character; and it seems to the eye to be laid on the disc of the Sun."*[74]

Thales founded the Milesian School of philosophy with his colleagues Anaximander and Anaximenes, who both came up with their own ideas on the nature of the world. Anaximenes thought that the cause of all things was air, the Earth was flat, and the fixed stars were attached to a revolving crystalline sphere. The idea of the heavens being some sort of transparent rigid sphere rotating about the Earth permeated astronomy for the next 2000 years. Anaximander thought the Moon was a circle far larger than the Earth, the Moon and Sun

---

[74] Aetius II.28, The Oxyrhynchus Papyri, 3710

were vent-like openings to a giant wheel of fire and the eclipses depended upon the viewing angle to the turning of the wheel. Empedocles of Akragas thought that the planets were fiery masses moving freely in space beyond the Moon. Heraclitus of Ephesus explained the Sun and Moon as bowls full of fire. As the Moon's bowl rotated it caused the phases, and eclipses were the result of a rotation of the convex side of the bowls to face the Earth. This may sound crazy to us today, but at least they were trying to make sense of the cosmos, and I doubt that I would have come up with anything better.

How I would love to have a conversation with Pythagoras, the philosopher who left such a mark in history. The enormous influence of Pythagoras led to Bertrand Russell naming him as the most important philosopher in history. Newton and Einstein both credited him for his profound influence. However, Walter Burkert, a German scholar of Greek mythology at the University of Zurich, argued that Pythagoras was a charismatic political and religious teacher, but that the number theory attributed to him was really an innovation by Philolaus, the successor of Pythagoras in the 5th century BC. According to Burkert, Pythagoras never dealt with numbers at all, let alone made any noteworthy contribution to mathematics. It might therefore be more appropriate to credit the Pythagorean school of thought, which lasted for two hundred years, spanning the 6th to the 4th centuries BC.

No original writings of Pythagoras remain, but the contributions from the Pythagoreans are mentioned by many later Greek authors. The central Pythagorean idea was that the world was understandable through numbers and harmony. Motivated by the regularities of the planetary motions and how the harmony of musical sounds depends on regular intervals, they asserted that the motions of the Sun, Moon and planets gave rise to musical tones that we cannot hear because we have heard them constantly from birth.

At this time in history, the Moon was considered to be a planet by most ancient astronomers. All the planets and the Sun were thought to orbit the Earth. It certainly seemed that way because they all crossed the sky following the same path. It is a bit of a coincidence that the Moon seems to move along the ecliptic, the orbital plane of the Earth and planets around the Sun. Most moons of other planets orbit the equator of their host planet, but our Moon does not. Its orbit about the Earth is tilted by as much as 28 degrees to Earth's equator. Because Earth itself is tilted to the ecliptic by 23.5 degrees, the Moon follows the ecliptic plane within five degrees. If the Moon orbited Earth's equator it would have a rather different path across the sky than the planets, and it may have been thought of as different from the other worlds in our solar system. Of all the planets, the Moon was considered the closest because its surface can be seen as a disk and not a distant point.

Pythagorean cosmology was based on the world being formed from the four elements: earth, water, air and fire. The Earth was imagined as spherical and inhabited all over, an incredible step forwards, but the Sun and Moon were still regarded as flat objects. The daily rotation of the starry heavens and the Sun were caused by the Earth being carried in 24 hours around a circular path orbiting a central fire, the home of Zeus. Although this gives the same qualitative picture as a spinning Earth, the Pythagoreans thought the

Earth was not spinning, motivated by the fact no other celestial object could be observed to be spinning – the Moon constantly faces the Earth. The fact that the central fire about which all celestial bodies moved had never been seen did not seem to concern the Pythagoreans – they thought it must be hidden or visible from the other side of our world.

The Pythagoreans thought ten was a perfect number, and since the heavenly bodies numbered nine (in order after the Earth; Moon, Sun, Venus, Mercury, Mars, Jupiter, Saturn and the slowly rotating rigid sphere containing the fixed stars) they believed there must be a tenth planet, a Counter-Earth. It was a bold step to remove the Earth from its position at rest at the centre of the universe. Some Pythagoreans thought that the Moon itself was a body like the Earth, hosting plants and animals. Others thought that the markings on the Moon were a reflection of Earth's land and seas on a perfectly smooth crystalline lunar surface, an idea maintained by Aristotle.

Parmenides of Elea was one of the founders of the Eleatic school of philosophy in the 5th century BC. Its members included Zeno of Elea and Melissus of Samos, who questioned the nature of reality, space and time. Some of their paradoxical thought experiments are still debated by philosophers today. Theophrastus, the successor to Aristotle, attributed Parmenides and not Pythagoras as the first to perceive that the Earth was spherical. He correctly stated that the Moon shines with light borrowed from the Sun. He thought the Sun and Moon formed from matter that was once part of the Milky Way, that the Sun was made from a *"hot and subtle substance"*, and the Moon from the *"dark and cold."* He assumed incorrectly, as did Anaximander, that the stars were closer than the Moon. A strange conclusion given that the occultation of bright stars by the dark part of the crescent Moon clearly reveals which is more distant.

Anaxagoras was an Ionian from the 5th century BC who lived in the neighbourhood of Smyrna in what today is Turkey. Like many of the ancient philosophers and scientists until the 20th century, he came from a rich family but gave up his wealth in order to devote himself to science. He stated that the Moon was an inhabited world, illuminated by the *"red hot stone that was the Sun"* and gave the correct explanation for lunar and solar eclipses as well as for the lunar phases. He must have therefore guessed that the Moon had a spherical form.

When there are gaps in scientific knowledge there is often great confusion. If the Moon shone with reflected light of the Sun, why could a faint light be seen on the unlit portion during a crescent Moon phase? The explanation of Earthshine, the reflected light of the Earth to the Moon and back had to wait until it was explained by Leonardo da Vinci in the 16th century. And why did the Moon glow blood red during a lunar eclipse? Since the Moon lies completely in the shadow of the Earth, it must have seemed that the Moon must absorb sunlight and radiate it like a glowing ember.

It is difficult to reconstruct the reasoning and thoughts of these philosophers because their original texts are lost to history, but we can see that they were boldly trying to understand the cosmos. Some of their ideas sound ridiculous to us today, but only with the knowledge that we now have. And some of their ideas were brilliant insights that turned out to be correct. The atomic theory of matter by Leucippus and Democritus in the 5th century BC was another great example. According to Aristotle, Democritus compared the Earth to a discus, he placed the Moon closer to the Earth than the Sun and thought that they were large solid masses but smaller than the Earth. He thought the markings on the face of the Moon were caused by the shadows of mountains and valleys, and that the Milky Way was actually the light from a multitude of distant stars – remarkably correct speculations that were confirmed 2000 years later with the invention of the telescope.

There was little agreement amongst the early Greek philosophers as to the nature of the Moon. Thales, Anaxagoras, Democritus and Plato all maintained that the Moon was Earth-like, but others argued that the Moon burned with a gentle fire like the Sun. Pythagoras and Aristotle thought it was a perfectly spherical crystalline mirror. In his famous 2nd century dialogue *'On the face which appears on the orb of the Moon'* Plutarch writes that the Moon cannot be made of a crystalline substance since solar eclipses would not be possible. And that the absence of a bright reflected image of the Sun and the Earth means it is not polished but must be more similar to our Earth.

Some of the interpretations of the cosmos from the classical Greek period could be immediately ruled out with a little foresight. A simple thought experiment should have convinced Aristotle that the Moon was not a mirror and its markings were not a reflection of the Earth – as the Moon moves around the Earth the reflected patterns should change with time, but the markings on the Moon never change. At this time in Greece there was not a long history of astronomical observations. Does this mean that the vast knowledge of the Babylonians had not yet reached them? It does seem that way when we look at the development of the early Greek calendar systems.

Meton of Athens, in the 5th century BC, was one of the first early Greeks to take careful astronomical observations. He developed a lunar-solar calendar based on the fact that 19 solar years are about equal to 235 lunar synodic months, a period relation that was also known to the Babylonians and the ancient Chinese. This means that during one 19-year period, a thirteenth month has to be intercalated seven times. Prior to this time, the Greeks had followed a lunar calendar that was aligned to the solar year using the solstices and equinoxes, but there was no apparent rule for including intercalary days or months – that was decided in an ad-hoc way city by city. To make it more confusing, there were usually several calendar systems in use at any one time. Athens adopted the Metonic cycle to align their lunar and solar calendars, which was fixed by the observations of Meton of the summer solstice in 432BC. A Metonic cycle lasts 6,940 days and only has an error of one day every 219 years because the lunar month and solar year do not align exactly.

A century later, the Greek astronomer Callippus made a more accurate calculation of the duration of a solar year at 365.25 days. The new improved lunar-solar calendar was a simple matter of multiplying the Metonic cycle by four and omitting one day from the last 19-year cycle. This gave another solar cycle of 76 years that consisted of 940 lunar cycles or 27,759 days. The world's oldest known astronomical calculator, the incredible Antikythera Mechanism dating from the 2$^{nd}$ century BC, performed calculations based on both the Metonic and Callipic calendar cycles, with separate mechanical dials for each.

At this time in Athens, around the 3$^{rd}$ century BC, there were five separate calendars that could be consulted: Olympiad, Seasonal, Civil, Conciliar, and Metonic – depending on what event needed to be kept track of. The primary need for a seasonal calendar, or parapegma, emerged because the ancient Greeks needed to know when the weather might change to regulate activities such as agriculture, navigation and warfare. The parapegmata were tables of observations that noted specific visible astronomical phenomena within a given year. Catalogued for centuries by various astronomers, parapegmata were recorded as a list of seasonally recurring weather changes in relation to the first and last appearances of stars and constellations, the phases of the Moon and the equinoxes and solstices. The famous 3$^{rd}$ century works of the Greek poet Aratus, '*Phenomena*' and '*Diosemeia*' contain poetic versions of the parapegma. A typical passage reads: *"But when the clear light from the stars is dimmed, though no thronging clouds veil, nor other darkness hide nor Moon obscure, but the stars on a sudden thus causelessly wax wan, hold that no more for sign of calm but look for storm."*

The next reform of the calendar system took place centuries later by Julius Caesar to remove the need for intercalary months. The Roman calendar had used 12 months of varying lengths such that the year was 355 days long. However, because they did not systematically include intercalary months, by the 1$^{st}$ century BC the existing Roman calendar had slipped three months with respect to the seasons. On the advice of the astronomer Sosigenes from Alexandria, Caesar introduced a new system, resetting the calendar by adding ninety days to the year 46BC, the so called 'year of confusion', and started a new calendar. The year was defined as 365 days long, and every four years (a leap year) an extra day was added to match the 365.25 day solar year.

**Plato's students**

Plato added little of value to our understanding of astronomy, and in one sense he set astronomy back by insisting that the lunar and planetary motions should follow circles, which were the perfect form.

Plato advised one of his students, Eudoxus of Cnidus who lived in the 4th century BC, to look into detail at the motions of the celestial bodies. Eudoxus travelled to Egypt and spent a year there learning of the Egyptian observations of the planetary and lunar motions. To explain these observations he came up with a rather sophisticated model in which the celestial bodies all rotated around the Earth, guided by rigid concentric crystalline spheres. In this model the Moon moved on a circular orbit about the Earth. But the primary guiding sphere itself had three additional spheres that tipped and tilted the Moon's orbit such that it matched the observations. The motions of the planets were also guided by additional spheres, which were an attempt to match their sometimes retrograde motions. It was all rather complex, but it seemed to reproduce the limited observations of the time. This was the real foundation of the celestial spheres, but it was a continuation of the ideas of Anaximenes. In Eudoxus' model there were three spheres each to guide the motions of the Moon and the Sun and four each for the five known planets, making 26 spheres in all.

Aristotle was another 4th century BC student of Plato. His books are an invaluable source of information from those who came before and from whom nothing survives. Much of his writings on physics and astronomy were incorrect, others were bold steps forward.

Aristotle made the following very interesting observation in his 350BC text '*On the Heavens*': He first argued that the natural equilibrium shape of a celestial body would be a sphere because of the force that always pulls things towards the centre. He then goes on to write: *"…the evidence of our eyes shows us that the Moon is spherical. For how else should the moon as it waxes and wanes show for the most part a crescent-shaped or gibbous figure, and only at one moment a half-moon? And astronomical arguments give further confirmation; for no other hypothesis accounts for the crescent shape of the Sun's eclipses. One, then, of the heavenly bodies being spherical, clearly the rest will be spherical also."*[75] Perhaps this correct explanation of the phases of the Moon was his own, or perhaps it was due to Anaxagoras.

Knowing that the Moon was a sphere, Aristotle also wondered why we only see one side of the Moon and not the far side. He assumed that the Moon had no ability to spin because it was carried in a circle within a rigid crystalline sphere.

---

[75] "On the Heavens", Aristotle 350BC, Part 11, translated by J. L. Stocks, 1922, Oxford Clarendon Press.

He reveals the influence of the Egyptians and Babylonians when discussing the evidence that the planets lie beyond the Moon: *"Yet these planets are farther from the centre and thus nearer to the primary body than they, as observation has itself revealed. For we have seen the Moon, half-full, pass beneath the planet Mars, which vanished on its shadow side and came forth by the bright and shining part. Similar accounts of other stars are given by the Egyptians and Babylonians, whose observations have been kept for very many years past, and from whom much of our evidence about particular stars is derived."*[76]

Aristotle also gives the known evidence for a spherical Earth: *"Again, our observations of the stars make it evident, not only that the Earth is circular, but also that it is a circle of no great size. For quite a small change of position to south or north causes a manifest alteration of the horizon. There is much change, I mean, in the stars which are overhead, and the stars seen are different, as one moves northward or southward [....] Also, those mathematicians who try to calculate the size of the Earth's circumference arrive at the figure 400,000 stadia. This indicates not only that the earth's mass is spherical in shape, but also that as compared with the stars it is not of great size."*[77]

Here Aristotle references those who had measured the size of the Earth, but he frustratingly does not mention how this was done or who did it. Possibly they used the varying angle to the stars as seen from different locations. The length of a Greek stadia is not well determined, but it is somewhere between 150 and 210 metres, which means the circumference

---

[76] Ibid, Part 12.
[77] Ibid, Part 14.

of the Earth that Aristotle quoted was between 60,000km and 84,000km, a factor of two too large.

A more accurate value was determined by Eratosthenes of Alexandria, who wrote a text on the size of the Earth towards the end of the 3rd century BC. Eratosthenes accomplished this without moving from his home. He had heard that on the day of the summer solstice the reflection of the Sun was visible at the bottom of a deep well in a southern city called Syene. That means that a vertical stick in Syene would cast no shadow because the Sun was exactly overhead that day. But at the same time at his home town of Alexandria the stick cast a shadow with an angle of just over seven degrees. This angle represents the difference of latitude of the two sites on the surface of a spherical Earth. Distances between cities were roughly known from traders who regularly made the journeys. The distance between Alexandria and Syene was 5000 stadia, thus the circumference of the Earth was 250,000 stadia. The value of Eratosthenes stadia was 157.5 metres, so Erathosthenes' circumference of the Earth was 39,690 kilometres, which is accurate to better than one percent!

From this point in history, there were practically no scientists who doubted the spherical shape of our planet or the Moon. Aristotle was of course influenced by earlier philosophers. Some of his observations were original, others from sources he did not cite. I wonder who 'those mathematicians' were who first tried to calculate the size of the Earth? Perhaps it was Eudoxus who observed the stars from both Egypt and Greece. That the Earth itself was spinning in space was not accepted until much later, but it had been proposed around the time of Aristotle.

Hicetas of Syracuse was one of the last Pythagorean philosophers in the 4th century BC. He was the first to suggest that the Earth itself was spinning on its axis, turning once in 24 hours, thus abandoning the idea of an orbit about a central fire. We know very little of Hicetas, just a few mentions including this most important reference from the Roman philosopher Cicero in the 1st century BC: *"Hicetas of Syracuse, as Theophrastus says, holds that the heaven, the Sun, the moon, the stars and in fact all things in the sky remain still, and nothing else in the universe moves, except the earth; but as the earth turns and twists about its axis with extreme swiftness, all the same results follow as if the earth were still and the heaven moved."*[78]

### Our place in the solar system

The Pythagoreans were perhaps the first philosophers to speculate as to the Earth's place in the solar system – they estimated distances to the celestial bodies based on their apparent times to orbit the Earth. Their theories on the harmony of math, geometry and music gave rise to the 'harmony of the spheres'. A romantic idea that captivated imaginations from Plato to Shakespeare until the end of the middle Ages as exemplified in William Shakespeare's *'Merchant of Venice'*:

*"There's not the smallest orb which thou behold'st*
*But in his motion like an angel sings,*
*Still quiring to the young-eyed cherubins:*
*Such harmony is in immortal souls;*

---

[78] "Aristarchus of Samos, the ancient Copernicus", 1913, Sir Thomas Heath, Oxford Clarendon Press, page 188.

*But, whilst this muddy vesture of decay*
*Doth grossly close it in, we cannot hear it."*

The first correct estimate of the distances to the Sun and Moon had to wait two centuries for Aristarchus of Samos. Little is known of the life of Aristarchus except that he spent at least some years at the museums and libraries in Alexandria. Aristarchus made several of the most notable contributions to astronomy in ancient Greece towards the end of the 3rd century BC.[79] The only surviving text is his work *'On the Sizes and Distances of the Sun and Moon'*.

First, Aristarchus calculated the relative distances and sizes of the Sun and the Moon. He noted that when the Moon was exactly half illuminated, the Earth, Moon and Sun would form a right-angled triangle with the Moon at the vertex of the right-angle. Aristarchus only needed to measure the angle between the Moon and Sun as viewed from Earth. He could then use the geometry of triangles developed by Euclid to determine the relative lengths of two sides of the triangle, the ratio of the Moon to Sun distance. Aristarchus measured 87 degrees, from which he inferred that the distance to the Sun was about 19 times the distance to the Moon. If you use the correct angle of 89.8 degrees then you would find that the Earth-Sun distance is actually 400 times the Earth-Moon distance. Aristarchus also reasoned that since the angular size of the Sun and Moon were the same (by using the fact that during a solar eclipse the Moon almost perfectly fits on top of the Sun), then simple geometry tells us that if the Sun is 400

---

[79] "Aristarchus's On the Sizes and Distances of the Sun and the Moon: Greek and Arabic Texts", 2007, J.L. Berggren and N. Sidoli, Archive for History of the Exact Sciences, 61, 3, 213.

times the distance of the Moon, the Sun must be 400 times the size of the Moon.

With a second method Aristarchus determined both the absolute sizes and distances to the Sun and Moon. He used the geometry of a lunar eclipse to calculate the radii of the Moon and Sun in terms of Earth radii. This second calculation involves a lot more triangles and geometry, but the basic idea is that he could estimate the size of the shadow of the Earth at the distance of the Moon and the size of the Moon relative to Earth's shadow. He could do this by timing how long it took the Moon to cross the shadow (the length of totality during a lunar eclipse) and knowing the orbital time of the Moon about the Earth.

Hipparchus and Ptolemy refined the estimates of Aristarchus, and the distance to the Moon was estimated to be around 60 Earth radii and its size 3.3 times smaller than the Earth. These were quite accurate estimates, and they were not improved upon until the invention of the telescope. Hipparchus and Ptolemy knew the Sun was over a hundred times the diameter of the Earth, which puzzles me as to how they thought this giant object moved around the Earth rather than the Earth moving around the Sun.

Another visionary idea of Aristarchus is preserved in the text '*Psammites*', also known as '*The Sand Reckoner*', written by his younger contemporary Archimedes. After describing the prevailing geocentric model he goes on to write that Aristarchus of Samos had published the hypothesis that it is the Sun and fixed stars that are immovable, and the Earth and planets move around the Sun. Aristarchus also knew that the fixed stars must be very distant since they showed no signs of parallax, apparent changes in position, when viewed from different places. Aristarchus correctly identified our place in the solar system and proposed the heliocentric model long before Copernicus in the 16th century AD.

Archimedes was certainly inspired by Aristarchus – his short text was one of the first 'popular science' works, in which he used the results of Aristarchus to calculate how many grains of sand it would take to fill the known universe! But aside from a few citations by other Greek philosophers, it seems the idea did not gather universal acceptance. One exception is the Chaldean astronomer Seleucus of Seleucia who supported and taught the heliocentric model of Aristarchus as well as the 24 hour rotation of the Earth.

Seleucus is known from the writings of Plutarch, Aetius, Strabo, and Muhammad ibn Zakariya al-Razi. The Greek geographer Strabo lists Seleucus as one of the four most influential astronomers from Hellenistic period, around 150BC. According to Plutarch, Seleucus even proved the heliocentric system through reasoning, though it is not known what arguments he used.

I find the lack of general acceptance of the ideas of Aristarchus somewhat puzzling. I can only imagine that it was either due to the ideas of Aristotle regarding the nature of motion, which he got completely wrong, or the notion that if the Earth were moving we would somehow feel it.

Even the last of the great Greek astronomers, Hipparchus of Nicaea, 2nd century BC, reverted to the geocentric model with a static non-rotating Earth at the centre of the universe.

Hipparchus was the most renowned of the Greek astronomers in the Hellenistic period, carrying out the most detailed observations of the motion of the Sun, Moon, stars and planets. Hipparchus developed astronomical tools such as the astrolabe that enabled him to measure accurate positions of the stars.

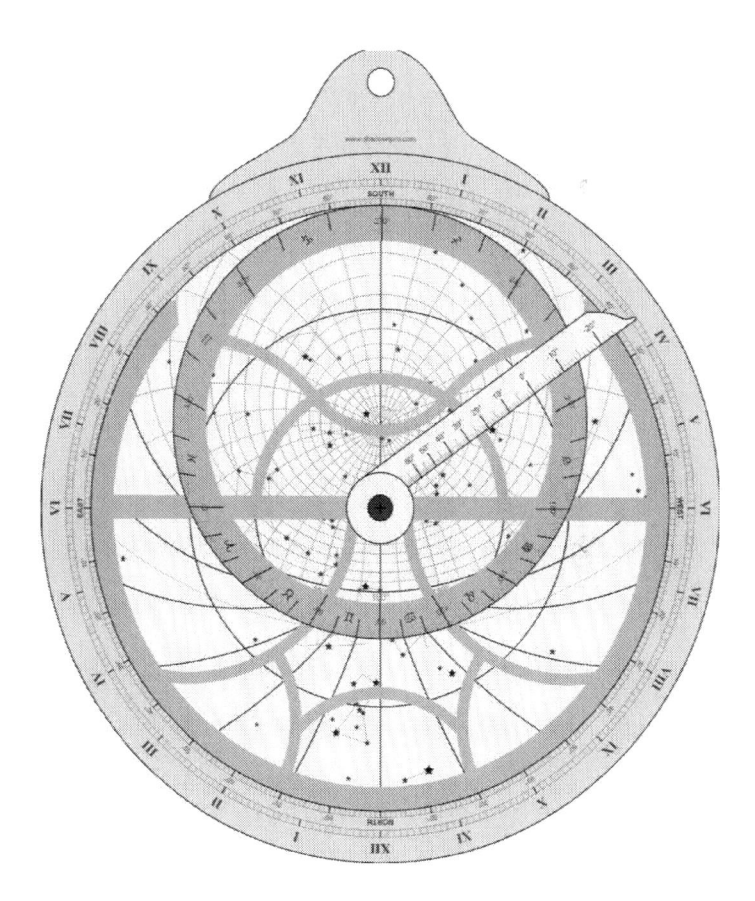

*Diagram of an astrolabe*

From the island of Rhodos, Hipparchus created the first extensive catalogue of nearly 1000 stars, with their coordinates measured to an accuracy of about one third of a degree. When he compared his measurements with those made 150 years earlier by the Alexandrian astronomer Timocharis, he found a systematic difference in the times when the stars reached their same ecliptic longitude.

Hipparchus correctly interpreted his findings as a continuous motion of the equinoxes along the ecliptic. Or in terms of the geocentric model, the celestial sphere was wobbling as it rotated around the Earth. The correct interpretation is that it's the Earth that is wobbling, rather like a wobbling spinning top that is about to fall over. The axis about which the Earth spins is precessing – the celestial pole slowly traces out a circle around the ecliptic pole about once every 26,000 years. Isaac Newton later showed that this precessing motion is caused mainly by our Moon.

## 9. Of Astrologers and Astronomers

Western astronomy began with the Sumerians, was refined by the Babylonians and used by the ancient Greeks to begin to develop a scientific understanding of the cosmos. The progress was remarkable, and if the pace of knowledge had been maintained we would be much further in our knowledge today. However, after the ancient Greeks there was a long period of scientific stagnation.

Attention turned to the art of astrology, which required a model for the accurate prediction of the planetary positions against the backdrop of the constellations. Two hundred and fifty years after Hipparchus came the last proponent and compounder of Greek astronomy, Claudius Ptolemy of Alexandria. He deserves a special mention, not only for his work '*Almagest*', but for his extensive text on astrology, the '*Tetrabiblos*'. Unfortunately, in my opinion, both of these works set science back for more than a millennium.

Ptolemy had access to the ancient scrolls of knowledge at the great libraries of Alexandria and wrote '*Almagest*', the only surviving comprehensive text on early western astronomy. It was originally titled 'Μαθηματικὴ Σύνταξις' (*Mathēmatikē Syntaxis*) in ancient Greek, but the surviving copies are thanks to the Arabic translations that preserved his work. It became one of the most influential scientific texts of all time, preserving the findings of many Greek philosophers whose original works were lost. It became widely translated, spread through the Arab world and reached as far as India.

Ptolemy further developed the crystalline shells model, adding even more imaginary circular motions that moved around even more imaginary points in space – over seventy spheres in total!

**Backwards steps**

The primary reason that Ptolemy constructed such a complex model for the motion of the celestial bodies was for his 'astronomical prognosis', which was the foundation of Western astrology. Ptolemy thought that the Sun, Moon, stars and planets all exerted an influence on life. And since their positions determined the strength of the various effects, Ptolemy needed an accurate way of predicting their past and future paths. The existing models of Aristotle and Hipparchus were not accurate enough for his purposes. The complex system of offset rotating spheres was not an attempt to determine the cause of the motions. There was no discussion of forces or physics in the Almagest. It was all developed for his next great text, the *'Tetrabiblos'*.

The Tetrabiblos, or four-part book of Ptolemy, continued to be the defining astrological text for the following 1000 years. Although Ptolemy does not deserve the entire credit, or blame, for developing astrology as we know it today, he attempted to turn all of the ancient ideas of parapegmata, omens and myths into something resembling a science. Many Greek philosophers in the Hellenistic period dabbled in astrology. And Plato is responsible for a large part of the shift from the ancient Greek philosophers who took the planets and stars to be material bodies of one substance or another. Plato advocated that the celestial bodies had a divine status and contained the 'nous', or soul. Unfortunately, Ptolemy took the ideas of Plato much further.

The Tetrabiblos is mish-mash of omens, myths, folklore and astronomy from the Sumerians, Babylonians, Egyptians and Greeks. The opening sentence sets the scene: *"That a certain power, derived from the æthereal nature, is diffused over and pervades the whole atmosphere of the earth, is clearly evident to all men."* Ptolemy goes on to mention the power of the Sun over the seasons and all living things. Next he mentions the influence of the Moon in two sentences that pervade folklore and myths until today: *"The Moon, being of all the heavenly bodies the nearest to the Earth, also dispenses much influence; and things animate and inanimate sympathise and vary with her. By the changes of her illumination, rivers swell and are reduced; the tides of the sea are ruled by her risings and settings; and plants and animals are expanded or collapsed, if not entirely at least partially, as she waxes or wanes."*

The strength and associated effects of the planets were determined by their distances from the Earth and Sun. Ptolemy's ordering of the celestial bodies was the same as Aristotle had proposed – the distances from Earth being related to the orbital times: Moon, Mercury, Venus, Sun, Mars, Jupiter, Saturn and the eighth sphere containing the fixed stars. Although the basic idea of relating orbital times to distance is correct, it only works if you have the planets orbiting the Sun. Ptolemy's ordering was incorrect because in his model they moved about the Earth. The foundations of modern astrology are based on this incorrect picture, with Venus placed further from the Earth than Mercury and closer to the Sun than Mercury, which I find slightly amusing.

In Babylonian times, astronomical events such as eclipses of the Moon and Sun were seen as omens from the gods. Although once the regular patterns in eclipses were discovered and the mechanism of eclipses understood by the 3rd century BC, I find it strange that Ptolemy and others went even further with the astrology of eclipses. In his text he writes about the importance of noting the time, duration and type of eclipse, the angle to the event, the position of the background stars and even the colour of the sky or haloes about the Sun and Moon. Several hundred pages of pure nonsense that became a revered book with the status of the bible.

As the centuries passed, belief in astrology grew and exerted its influence on mankind, arousing fear and hope but always doomed to disappoint. Astronomy did not significantly advance after Hipparchus. During the Roman Empire astronomy was performed just to keep track of the constellations and planets, and to look for the omens brought by eclipses and comets. In mediaeval times the production of horoscopes became a big business, and the sales of almanacs sored. Astrologers held positions of importance and were consulted before any decisions of importance were made. Universities created chairs of astrology, physicians adopted it for their medical practises, kings and queens ruled their countries with it.

Little had changed by the 17th century as the following text from 1676 reveals: *"Good to purge with electuaries, the moon in Cancer; with pills, the moon in Pisces; with potions, the moon in Virgo. Good to take vomits, the moon being in Taurus, Virgo, or the latter part of Sagittarius; to purge the head by sneezing, the moon being in Cancer, Leo, or Virgo; to stop fluxes and rheums, the moon being in Taurus, Virgo, or Capricorn; to bathe when the moon is in Libra, Aquarius, or Pisces; to cut the hair off the head or beard when the moon is in Libra, Sagittarius, Aquarius, or Pisces."*[80]

Even the great 16th and 17th century scientists such as Tycho Brahe, Galileo Galilei and Johannes Kepler all held astrology in high esteem. But by the end of the 17th century, the second scientific revolution was well underway, and the influence of astrology waned. Not a single word on astrology, either for or against it, is to be found in the works of famous scientists such as Christiaan Huygens or Isaac Newton.

Although much progress was made by the ancient Greeks who were on the edge of an industrial revolution, that all came to a dismal end.[81] Why then is there so little for me to report from the 1500 years following Hipparchus? Progress in astronomy came to a standstill, but religion and astrology began to flourish and spread. Until the 15th century there were no new insights or progress made towards understanding the nature of our Moon nor on the workings of the solar system.

In the East the knowledge of the Greeks was preserved and copied, but was not taken significantly further. Did scientific discovery have to pause until the refinements in glass making led to the invention of the telescope? The human mind alone

---

[80] "The knowledge of Things Unknown: The Husband-Man's practise. Or Prognostication for ever. As teacheth Albert, Alkind, Haly and Ptolomey. With the Sheperds perpetual Prognostication for the Weather", 1676, London, pages 108/109.

[81] Thanks to the Romans... perhaps.

could have taken the existing observations and mathematics and continued the path of discovery.

But the knowledge of the Greeks never completely faded. Ancient manuscripts were kept hidden in monasteries, and monks had plenty of time to study them. There were glimmers of light from gloomy places, such as came from the extensive writings of the Venerable Bede around 700AD in Northumbria, my birthplace. In the peaceful monasteries of the North of England, he studied old texts originating from the Greek philosophers and reinforced their views of a spherical Earth and the motions of the planets. In the 12th and 13th centuries, Arabic translations of Aristotle came to Europe. Roger Bacon wrote his *'Opus Majus'* in 1267 and began a reform in natural philosophy – eloquently writing on how the scientific method is the way to make progress.

Arabian, Indian and Chinese astronomy continued independent of events in Europe. There was less religious persecution, and science could have progressed unhindered. But despite advances in mathematics and the natural sciences, even the greatest astronomers of India and Arabia added little knowledge to our understanding of the Moon, let alone the cosmos beyond. Until as late as the 17th century, Chinese philosophers thought the world was flat and square, a design reflected in early Chinese architecture.

However, the art of astronomy was kept alive. The ancient texts had been studied and many of the observations made by the Babylonians and Greeks were improved upon. There were advances in astronomical instruments and mechanical timekeeping devices. Great observatories were constructed, ever more detailed observations of the night's sky were made and the precision of astronomical measurement increased.

## The Renaissance and the rebirth of Astronomy

By the 13th century there was a renewed interest in learning Greek throughout Europe. Ancient manuscripts were sought after, translated and studied. A particularly influential book *'De sphaera mundi'* by Johannes de Sacrobosco in 1230, collated the works of Ptolemy and Islamic astronomers. It was widely used in universities and copied by hand until the European invention of the printing press in the 15th century. Aristotle's writings were particularly influential and were widely debated by early Renaissance scholars. The philosophical thoughts on the Moon and planets from this time were rather like the ideas that had been made a thousand years earlier in the classical Greek period.

The idea of a mirror Moon reflecting the Earth's surface was discussed by Albertus Magnus in the 13th century, who argued that the features on the Moon may be due to the absorption of light beneath its surface. Robertus Anglicus said that the brighter and darker features on the Moon could be due to different materials on the surface of the Moon, arguing that viewed from afar, Earth's oceans would appear brighter and the land darker. Others argued the opposite, such as the French scholars Jean Buridan and Nicole d'Orsesme. And some thought the dark patches on the Moon were clouds. Albert of Saxony debated the light from the Moon, using the faint glow of the dark area of the Moon during its crescent phase to argue that it soaked up the light of the Sun but also had an intrinsic light of its own.

Of all the early Renaissance philosophers, there is one in particular who stands out – Leonardo da Vinci. In 2019, the publication year of this book, it is the 500th anniversary since the death of Leonardo and the 50th anniversary of the first humans to walk on the Moon. Leonardo was a true polymath – an inventor, palaeontologist, mathematician, musician, sculptor and painter and far more. But relevant for this text on the discovery of our Moon, Leonardo was also fascinated by astronomy.[82]

He commented on the thoughts of the early renaissance philosophers: *"some have said that vapours are given off from the Moon after the manner of clouds, and are interposed between the Moon and our eyes. If this were the case these spots would never be fixed either as to position or shape; and when the Moon was seen from different points, even although these spots did not alter their position, they would change their shape."* [83]

Most of Leonardo's insights were never published, but he left a record of his thoughts in his extensive notebooks. He correctly speculated as to the nature of the Sun: *"The Sun has substance, shape, movement, radiance, heat and generative power; and these qualities all emanate from itself without its diminution."*[84] And he rejected the Ptolemaic Earth-centred universe: *"The Earth is not in the center of the Sun's orbit, nor at the center of the*

---

[82] "Leonardo da Vinci: A Review", 2008, K.H. Veltman, Leonardo, 41, 4, 381.

[83] "The notebooks of Leonardo da Vinci", 1955, E. MacCurdy, pub. George Braziller, New York, first published 1939 by Renal and Hitchcock Inc. page 285.

[84] Ibid. page 274.

universe, but in the center of its companion elements, and united with them"* [85] and he wrote *"The Sun does not move."*[86]

He understood gravity as a force that held the oceans fixed onto our spinning world and its universal nature: *"Gravity is limited to the elements of water and earth; but this force is unlimited, and by it infinite worlds might be moved if instruments could be made by which the force could be generated."*[87] He certainly had the knowledge to build a telescope, but it is not known if he did. Leonardo wrote: *"It is possible to find means by which the eye shall not see remote objects as much diminished as in natural perspective... and so the moon will be seen larger and its spots of a more defined form."*[88]

Leonardo is most famous for his few defining works of art, but perhaps less people know that he was also the first to accurately sketch the features on the surface of the Moon. In his notebook, the *'Codex Atlanticus'*, there are two pen and ink sketches of the Moon made between 1505 and 1508 in Italy. They are about two centimetres across, and one can recognise the features that make up 'the Man in the Moon'. Elsewhere, Leonardo sketches the Moon in black and white chalk and shows the western half of the Moon. It is about 17 centimetres in diameter and was drawn around the year 1513. It seems like the eastern part of the Moon was drawn on a page that is now missing. The major lunar maria can clearly be recognised. There is a circular mark in the lower right hand

---

[85] Ibid. page 281.
[86] Ibid. page 293.
[87] "The Notebooks of Leonardo da Vinci", 1888, Translated by Jean Paul Richter, project Gutenberg eBook, page 859.
[88] Ibid. page 869.

sector of his drawing, which could be the location of the Tycho crater – this crater is invisible to the naked eye, but a small telescope would have revealed its presence.

In several of his notebooks he wrote about the Moon and astronomy at length. *"If you keep the details of the spots of the Moon under observation, you will often find great variation in them, and this myself have proved by drawing them...."*[89] He argues that the spots are on the Moon itself: *"Others say that the surface of the Moon is smooth and polished and that, like a mirror, it reflects in itself the image of our Earth. This view is also false, inasmuch as the land, where it is not covered with water, presents various aspects and forms. Hence when the Moon is in the east it would reflect different sports from those it show when it is above us or in the west...A second reason is that an object reflected in a convex body takes up but a small portion of that body, as is proved in perspective."*[90] In another better known sketch in his 'Codex Hammer', Leonardo draws the thin crescent Moon and the ash-grey dim light that he correctly identified as earthshine – the light from the Earth reflected onto the Moon and back to our eyes. And he wrote *"If you were standing on the Moon or on a star, our Earth would seem to reflect the Sun as the Moon does."*[91]

---

[89] "The notebooks of Leonardo da Vinci", 1955, E. MacCurdy, George Braziller, New York, first published 1939 by Renal and Hitchcock Inc. page 290.
[90] Ibid, page 286.
[91] "The Notebooks of Leonardo da Vinci", 1888, Translated by Jean Paul Richter, Project Gutenberg eBook, page 893.

*Two pages from Leonardo da Vinci's Codex Hammer with Earthshine described on the right.*

Science proceeds step by step, building upon the findings of generations before. Of course, some steps are bigger than others, and at rare times there can be giant leaps. The first bold steps were made by Thales, followed by many inspirational ideas from the early Greek philosophers. Leonardo da Vinci was leaps ahead of his Renaissance contemporaries. Now the second scientific revolution was underway, and the time was right for Nicolaus Copernicus to show mathematically that everything was much simpler if our Moon orbited a spinning Earth, and the Moon and Earth together orbited the Sun once per year.

## Breaking the crystalline shells

The crystalline shells that were thought to carry the planets were shattered by astronomer Tycho Brahe. He systematically observed the position of the Moon and planets every night and recorded their positions using an astrolabe. He also observed a prominent comet in 1577 and showed that it had no parallax against the distant stars. He correctly concluded it must be beyond the orbit of the Moon, which went against the prevailing ideas throughout the Middle Ages that comets were an atmospheric phenomenon. Brahe also reconstructed the orbit of the comet and showed that it passed through the impenetrable homocentric crystalline spheres of Ptolemy, destroying the reality of the solid orbs that were thought to move the Moon and planets. Comets originate from the outer solar system and travel inwards on highly eccentric elliptical orbits. Brahe also noted that the comet's motion seemed to deviate strongly from a circle, more like an oval – the first departure from Plato's natural circles.

The German astronomer Johannes Kepler used the observations of Tycho Brahe to find patterns and relationships of the planetary and lunar orbits. He was also hoping that the correct model would naturally account for the all the small observational discrepancies that plagued Ptolemy and Copernicus's models of circular motions. At the beginning of the 17th century, Kepler came up with his three famous laws describing planetary motion. He showed that the planets moved on elliptical orbits with the Sun at their foci point. Finally, the break from the Moon and planets moving on circles! Kepler revealed the solar system in all its simplicity, reducing the number of parameters to just the eccentricity of the elliptical orbit and the distance from the Sun. But he struggled with explaining the complex details of the orbit of the Moon.

Although Kepler thought extensively about gravity and motion, and his ideas were more advanced than any before, he did not connect gravity with the force that kept the planets in orbit. Motivated by the study of Earth's magnetic field by the English astronomer William Gilbert, Kepler tried to explain the motions of the planets as if being carried around the Sun by a magnetic force. He also attributed the ocean tides to the magnetic-like attractive force from the Sun and Moon.

As the ideas of the astronomers were spreading, the church became wary of the new astronomy that disagreed with an immovable Earth at the centre of the universe described in the scriptures. In 1616, the consultors of the Inquisition at Rome declared the doctrine of the Earth's motion to be heretical. The Sacred Congregation banned the book of Copernicus and all related commentaries.

The advances in our understanding made by Copernicus, Brahe and Kepler could have followed quickly from the ancient Greeks. Even the developments on the fundamental aspects of motion from the great minds of Rene Descartes and Robert Hooke that led to the grand work of Isaac Newton could have taken place a millennia before. Little new observational data was needed by any of these 16th and 17th century researchers. With the observations made by the Babylonians, thought and mathematics alone would have been enough to come up with the heliocentric model, elliptical orbits and Newton's laws of motion and gravity.

### Seeing what others had never seen before

Perhaps the most important breakthrough was the invention of the telescope which revolutionised astronomy and biology, physics and chemistry. I think it is the most important invention ever made. The basic principle is to use one lens to capture more light than the eye can see and another lens to focus and magnify the image. This allowed astronomers to discover the universe of galaxies beyond our own and reveal our origins in the big bang. By changing the distance between the lenses, a telescope becomes a microscope, resolving objects smaller than the eye can see.

The history of the invention of the telescope is a little murky. Despite single magnifying lenses being used by the ancient Greeks and the speculations about optics by Leonardo da Vinci, the first patent for a telescope that used two lenses was filed by the Dutch optician Hans Lippershey in 1608. As soon as news of the telescope spread, many were constructed and pointed at the Moon. Galileo is credited as the first in 1609, but that is because he was quick to publish – even today, being 'second' with an idea relegates you into obscurity. In fact, the English scientist Thomas Harriot had observed and drawn the Moon a few months before Galileo, but his manuscripts remained unknown until 1784, and his lunar image was not published until 1965.

In March 1610, Galileo wrote his famous *'Sidereus Nuncius'* which described his first observations of the surface of the Moon, the vast number of stars in the Milky Way that could not be seen by eye and four orbiting moons of Jupiter. That removed one of the weaknesses in the heliocentric model –

that all the planets should orbit the Sun whilst our Moon was the only celestial object that orbited the Earth and not the Sun.

Now the status of our Moon shrank even further – thanks to Galileo it became one of five known satellites in our solar system! However, it may be the case that our Moon reverts to the status of planet, along with over 100 other solar system bodies, if a 2017 proposal to the International Astronomical Union is eventually approved. The complicated definition of planet that was approved by vote in 2011 included the criteria that a planet must have cleared the neighbourhood around its orbit. This is what demoted Pluto from a planet, since there are other similar sized bodies in Pluto's vicinity. I agree with the simple proposal to call everything a planet that is spherical and isn't a star. This means that it is massive enough for its gravity to have pulled its material into a spherical shape, but not too massive that it begins nuclear fusion. I will be voting yes for our planet Moon!

The German astronomer Simon Marius claimed to have discovered four of Jupiter's moons before Galileo. It turns out that indeed Marius independently made the discovery, and the date on his notes was eight days earlier than Galileo. However, Marius was using a different calendar system to Galileo.[92] Marius was using the older Julian calendar while

---

[92] By the 16th century, it was realised that the Julian calendar was using a solar year that was 11 minutes and 14 seconds too short. It doesn't sound like much, but after many centuries there was a ten day error in yearly events such as the equinox. Pope Gregory XIII employed the astronomer Christopher Clavius to find a solution. Since the error amounts to three days in 400 years, he suggested that century years (those ending in '00') should only be leap years if divisible by 400, which eliminates three leap years in every four centuries. The result is that 1600

Galileo was using the more recent Gregorian calendar, which was 10 days ahead. So Galileo actually made his discovery just before Marius.

Galileo's first observations revealed that the Moon's surface is covered with irregular features. Upon constructing an even more powerful telescope that could magnify by a factor of thirty, he said that the lunar features were mainly circular in shape, forming rings around depressed regions – what we now know are due to asteroid impacts. Galileo compared them to the 'eyes' in a peacock's tail. The poet John Milton visited Galileo in 1638 and in Book I of *'Paradise Lost'* (1667) compares the shield of Satan to:

*"the moon, whose orb*
*Through optic glass the Tuscan artist views*
*At evening, from the top of Fesole,*
*Or in Val d'Arno, to descry new lands,*
*Rivers, or mountains on her spotty globe."*

Galileo noticed the bright points of light at the terminator (the border between the light and dark part of the Moon) of the crescent Moon and recognised that these points were the tops of mountains illuminated by sunlight, whilst the surrounding valleys were very dark. The fact that the Moon is literally black or white was unlike the sunrise illuminating

---

and 2000 are normal leap years with an extra day in February, but the intervening 1700, 1800 and 1900 are not. The so-called Gregorian calendar began in 1582 and is the calendar that was ultimately adopted by the majority of the world. Over time, even the Gregorian calendar will slip with respect to the seasons. That is due to a lengthening of the day because of the influence of the Moon on the spin of the Earth. We will learn more about this fascinating effect in Chapter 12.

mountains on Earth. That's because the Moon has no atmosphere whereas Earth's atmosphere scatters some light into the valleys creating grey intermediate shadows.

From the length of the shadows from mountains as the sunlight began to cross the Moon Galileo measured that some of the peaks on the Moon were eight kilometres high. Although this was an overestimate, the highest lunar peaks on the giant crater rim of Mare Imbrium are about 5.5 kilometres high from base to peak. Galileo's finding was greatly at odds with the prevailing Pythagorean and Aristotelian philosophy that the Moon was a perfect sphere. Some still argued that there must be a perfectly smooth transparent crystalline shell covering the surface. In response Galileo said: *"I will grant this provided that, with equal courtesy, I be allowed to say that the crystal has on its outer surface a large number of huge mountains, that are thirty times as high as the terrestrial ones, and invisible because they are diaphanous."*[93]

There was more to come in Galileo's dismantling of Aristotelian cosmology. Venus is a planet about the size of the Earth, but at its distance it has an angular size of about one sixtieth of a degree – right at the limit of resolution of the human eye. Galileo also turned his telescope towards Venus in the summer of 1610 and saw that it was half illuminated – he found that Venus passed through phases rather like the Moon! Over the next months as Venus moved closer to the Sun, the illuminated part grew larger until Venus became a

---

[93] "Galileo in Rome: The Rise and Fall of a Troublesome Genius", 2003, W.R. Shea & M. Artigas, Oxford University Press.

full disk of light. Galileo discovered that Venus was also a spherical object illuminated by the light of the Sun.

When the Moon is close to the Sun, it lies between the Earth and the Sun and is only partly illuminated as a crescent. But because Venus orbits around the Sun, it reaches full illumination when it appears closest in the sky to the Sun. That's because Venus is orbiting beyond the Sun and reflects all its light towards Earth. Galileo observed this with his telescope, showing that Venus orbits the Sun and not the Earth, verifying the heliocentric model of Copernicus. If Venus had been a larger or closer world then the ancient Greeks could have proven the heliocentric model 2000 years ago because they would have witnessed its change of phases by eye.

In 1633 Galileo was punished for violating the orders of 1616 by teaching the heliocentric model. He was placed under house arrest at his villa in Arcetri near Florence until his death in 1642. However, news of Galileo's observations of the Moon spread rapidly, and there was a widespread interest in the ability to gaze in detail at this new world. It created more excitement than the 'discovery' of America by Christopher Columbus a century earlier. Ever larger telescopes were constructed and focussed on the Moon to map its surface and search for signs of life.

## 10. Observing the New World

The ancient Greek philosophers were doing what philosophers do. They began to ask the interesting questions and discuss possible answers. At that time, astronomy was not a science, but a branch of philosophy. Only when a subject has developed and becomes quantitative and predictive does it move from the realm of philosophy to become a scientific discipline of its own. In my opinion, it was the invention of the telescope and those 17th century philosophers like Descartes, Galileo and Newton that turned astronomy into its own scientific discipline. And like many sciences, astronomy had two branches.

There are the theoretical astronomers like Isaac Newton and Pierre-Simon Laplace, who tried to understand the workings of the solar system with mathematics and physics – they would be called astrophysicists or 'theorists' today. They studied in ever more detail the complex gravitational interactions between the three-body system of the Earth, Moon and Sun, derived the theory behind the generation of the tides and started to think seriously about the origin of the Moon.

The second branch of astronomers were those who developed the astronomical instruments and devoted themselves to observing the cosmos – that branch became observational astronomy, and today the practitioners are often called 'observers'. Galileo Galilei and Thomas Harriot could be called the first 'observational astronomers'. There has always been a strong collaboration between the 'observers' and 'theorists'. New observations give the theorists much to ponder, and new theories give the observers motivation to develop new instrumentation to look deeper at the universe.

Sometimes an observational astronomer has a good grasp of theoretical physics and can make great insights into astrophysics. Likewise, a theorist sometimes leads grand observational programs. And sometimes it can all go wrong, like the time I was left in control of a large and expensive telescope as I mentioned in the foreword. Or the amusing event from 2018 when a distinguished theorist colleague, who had a passion for gazing through his own small telescope, sent an urgent telegram to alert the world's astronomical community to announce a new bright transient object he had seen in the ecliptic plane. The following day a second rather apologetic telegram was sent to the world's great observatories whereby he announced his mistake and confessed the bright new object he had seen was actually the planet Mars!

And so with the advent of the telescope, astronomy branched out into these broad but entwined paths. The observers studied the Moon with their telescopes, mapping it in ever greater detail. It was as if a new world had been discovered, and the observers were like cartographers, peering in ever more detail at the landscape that the face of the Moon presented. They were searching for clues as to what the surface of the Moon was like. Did it host rivers and oceans or even life? Was there an atmosphere and weather? It was not until the end of the 19th century that telescopes improved sufficiently for the observers to study the surfaces of other worlds.

## Mapping the Moon

Whenever I show the Moon to someone through my telescope for the first time, there are always exclamations of excitement and wonder. And the excitement must have been even greater in the 17th century when news spread about the first telescopic observations of the Moon. An industry developed, constructing larger and more powerful telescopes, exploring the surface of the Moon like an explorer maps a new world. The next two centuries of observational lunar research were all about studying the surface of the Moon and making Moon maps in ever greater detail. This study became known as selenography.

The earliest known drawing of the Moon that shows any recognisable features is Van Eyck's *'The Crucifixion'*, painted around the year 1440. If you look at it side by side with a photograph of the Moon, then it shows the lunar markings that correspond to the features on the Moon. After Leonardo da Vinci, the third and last attempt to draw an accurate map of the Moon as seen by the unaided eye was made by William Gilbert at the end of the 16th century. His chart shows the outlines of dark and light patches on the Moon's face. It was published in 1651 in an unfinished work, *'De Mundo Nostro Sublunari Philosophia Nova'*, almost fifty years after his death.[94]

Gilbert sketched his lunar map for the fascinating reason described in his book – he regretted that no one did the same in antiquity so that he could determine if the face of the Moon had changed over the previous two thousand years. Contrary to most of his contemporaries, Gilbert believed that the light

---

[94] "The earliest maps of the Moon", 1969, Z. Kopal, The Moon, 1, 1, 59.

spots on the Moon were water, and the dark spots land. Gilbert also proposed the idea that a vacuum exists in the space between the Earth and the Moon. At this time it was commonly believed that the Earth and Moon shared an atmosphere and that the motion of the Moon through this shared atmosphere caused the ocean tides. If indeed the atmosphere of Earth reached to the Moon, the drag force from the Moon's motion would have caused it to spiral into our planet long ago.

By eye the Moon is covered with patches of different brightness. But at the distance of the Moon our eye can only resolve features that are larger than about 200 kilometres across. Galileo's first telescope gave an improvement in resolution of about twenty times that of the eye alone. From the shadows of sunlight within the depths of the lunar craters, Galileo recognised that these features were depressions, not mountains. He observed that many of the crater floors were covered with darker material, and that some had central peaks. He sketched the first detailed maps of the lunar surface in his famous text '*Sidereus Nuncius*' in 1610. It also contains his findings that the fuzzy Milky Way resolved into countless stars with his telescope – thus confirming the speculations of Democritus two millennia earlier. A first edition of Galileo's book was sold at auction in 2010 for $662,500, and some have tried to sell intricately produced fake copies for far more.

Around this time the large dark patches on the Moon were named maria (mare, singular), Latin for 'seas', since many astronomers thought these patches were giant oceans of water. They are the features that were dubbed the 'Man in the Moon' or the 'Rabbit in the Moon' by early cultures. However,

Galileo quite rationally wrote: *"Thus the Moon's spots could have been caused by enormous woods, but they could also have resulted from seas, if these were to have existed on its surface. And finally nothing is to prevent these spots from being a darker colour than the rest of the Moon's surface."*[95]

Galileo's sketches of the Moon are not particularly accurate, and the major features can barely be recognised (perhaps because Galileo was always in a hurry to get his results published), whereas Gilbert's map drawn with the naked eye shows the Mare Imbrium and a fuzzy merged view of several other lunar Mare. The year after Galileo's maps were published, Johannes Kepler developed the telescope further using a convex lens as an eyepiece, which allowed him to see the entire Moon at once through his telescope. There followed generations of observational astronomers who spent a large part of their lives studying the Moon, creating ever more detailed maps and speculating as to what lay on its surface.

---

[95] "The enigmatic face of the Moon", 2011, C. D. Galles and C. J. Gallagher, ASP Conference Series, Vol 441.

*Galileo's sketches of the Moon from 'Sidereus Nuncius'*

The greater the curvature in the lens, the closer to the lens the light from distant objects is focussed. But at this time it was difficult to accurately polish large curved lenses, and the early lenses were only slightly curved, so the focus point was a great distance from the primary lens. The astronomer Johannes Hevelius from Gdansk used a telescope where the front lens was nearly fifty metres away from his eyepiece! The entire wire and wood tube with its lenses was suspended from a tall wooden pole. Since it takes about an hour for the Moon to move a distance equal to its diameter, the telescope could be moved by hand to follow the Moon as it moved.

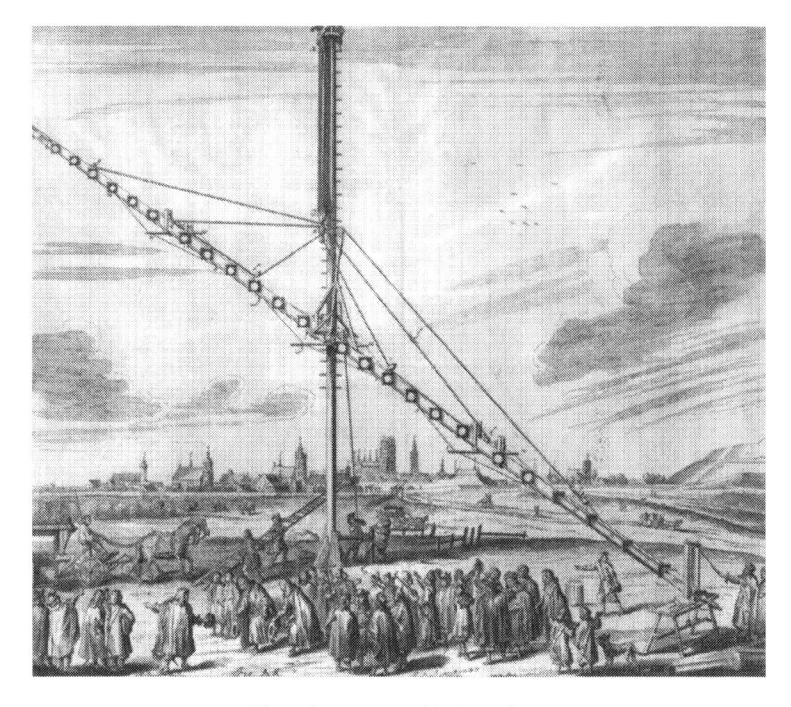

*The telescope used by Hevelius*

Hevelius constructed one of the first detailed charts of the Moon's surface in 1647. He believed in the existence of analogous regions to Earth, and named mountains, deserts, marshes, seas, lakes, islands, bays and straits. Some of his names were chosen for similar features – to his eye the great crater that we now call Copernicus was named Mount Etna. Hevelius engraved his chart onto metal himself and named it *Selenographia*. He also noticed that spots near the eastern and western edges of the disk of the Moon moved slightly and appeared to come in and out of view – he had detected the libration in longitude, an effect that allows us to see beyond the edge of the face of the Moon.

We only ever see one side of the Moon because the Moon is in synchronous rotation with its orbit, meaning that it spins around exactly once during one orbit of our Earth. That is not a coincidence, but a result of one of the most remarkable gravitational phenomena called 'tidal locking' that I will explain in Chapter 12. However, we can see a little more than half the surface of the Moon because it has a monthly cyclical apparent wobble which reveals part of the far side.

*Hevelius' map of the Moon*

The Moon is not moving around the Earth at a constant speed – during its elliptical orbit it speeds up as it moves closer to the Earth. But because the Moon rotates at a constant rate, this allows us to see an extra eight degrees of longitude of its eastern side. And as the Moon moves away from the Earth, its rotation is faster than its orbital motion, and this reveals eight degrees of longitude of its western side. This is referred to as longitudinal libration.

There is also an apparent wobbling motion of the lunar latitude. Because the Moon does not orbit Earth's equator, the rotational axis of the Moon appears to rotate towards and away from Earth during one complete orbit. This is referred to as latitudinal libration, which allows us to see almost seven degrees of latitude beyond the pole on the far side. We can also see an extra one degree of lunar longitude and latitude by viewing the Moon from different locations on Earth.

Thanks to this apparent wobbling motion of the Moon over a lunar month and by studying the Moon from different locations on Earth lunar cartographers were able to map almost 60 percent of its surface. This is why Moon globes produced prior to the Soviet mission around the Moon show a blank patch on their back that spans about 40 percent of the globe.

The lunar map of Hevelius was followed by that of the Italian astronomer and priest Giovanni Riccioli in 1651. Riccioli named the craters in honour of famous scientists and philosophers, and he named a particularly large crater after himself! His map was not as good as that of Hevelius, but his naming scheme was better, and so he made his place in history. Of the two hundred names on Riccioli's map, most are still in use today.

### The search for life on the Moon

John Wilkins was an English natural philosopher and one of the founders of The Royal Society, the oldest national scientific institute in the world. In 1640 he published his two volume work '*A Discourse Concerning a New World and another Planet*'. In the first book Wilkins proposed the Moon was inhabited by beings called the Selenites and described what he thought living on the Moon would be like. He also predicted that one day humans could fly to the Moon and live there.

I found Wilkins' books fascinating to read as they reveal the struggle between religion and astronomy – an attempt to come to terms with the cosmos with the limited understanding that was available at that time. The second book reveals this struggle in depth. Wilkins tries to reconcile the teachings of the church, that the world was flat and at the centre of everything, orbited by the Sun and planets. I particularly like this amusing quote: *"[Astronomy] is one of the most excellent Sciences in Nature, it may best deserve the industry of Man, who is one of the best Works of Nature. Other creatures were made with their Heads and Eyes turned downwards: would you know why man was not created so too? why it was, that he might be an Astronomer."*

The well-known Dutch scientist Christiaan Huygens ground his own glass telescope lenses and developed the first micrometer for measuring distances between objects viewed through his telescope. He discovered several new features of the Moon, as well as Saturn's rings and its largest moon, Titan. In 1695 Huygens wrote the first text describing from a scientific perspective what life could be like on the planets. *'Cosmotheoros'* was his last work, published after his death in 1698. Huygens discusses in depth the possible nature and behaviour of life on the planets, but he is sceptical about the presence of life on the Moon. He writes that Kepler thought the round craters on the Moon must be the work of lunar inhabitants, but argues that they are so large they must be due to natural causes.

Huygens goes on to describe his own observations of the Moon: *"Nor can I find any thing like Sea there, tho he and many others are of the contrary opinion I know. For those vast Countries which appear darker than the other, commonly taken for and call'd by the names of Seas, are discover'd with a good long Telescope, to be full of little round Cavities; whose Shadow falling within themselves, makes them appear of that colour."* He describes the lack of rivers on the Moon, that there must be no atmosphere because of the sharp rim of the Moon and the way in which stars suddenly blink out as they are occulted by the Moon. He argues that there are no clouds on the Moon because he never saw changes in the brightness of the surface features. Huygens was absolutely correct in his observations and in his concluding words wrote *"But as 'tis, I cannot imagine how any Plants or Animals, whose whole nourishment comes from liquid Bodies, can thrive in a dry, waterless, parch'd Soil."*

By the mid-18th century there were further improvements in telescope design. Achromatic refractor telescopes were invented with an additional convex lens that focused light of different colours correctly. Isaac Newton had perfected the use of mirrors rather than glass lenses. These so-called 'Newtonian' telescopes used a large curved mirror to gather and reflect light. They are easier to make and lighter than glass, although they do not provide quite as sharp a view as a refracting lens.

These developments were employed by the German astronomer Tobias Mayer to create a stunning orthographic lunar map, published posthumously in 1775. A few years later the German astronomer Johann Schroeter took mapping the Moon even more seriously and built his own private observatory to study the Moon. Rather like Percival Lowell a century later, Schroeter was intrigued with the idea that the Moon may be inhabited. He spent night after night studying its surface and created yet another detailed map of its features.

At this time it was widely believed by the most distinguished astronomers and the public that the Moon and the planets probably harboured some form of life. Perhaps the most famous astronomer of the 18th century was the German/British astronomer Sir William Herschel, who discovered the planet Uranus in 1781. He was an ardent supporter of Schroeter. Herschel considered the habitability of the Moon *"an absolute certainty"* and claimed some of the features on the Moon were artificially constructed. He even thought it possible that beneath the Sun's hot surface, there existed a cooler region where beings could live.

In Herschel's unpublished notes it is clear that he was searching for signs of life on the Moon with the world's largest telescopes at the time, with mirrors approaching one metre in size. In 1776 he claimed to see large areas of plant life on the Moon, and he searched for signs of lunar cities hidden amongst the lunar forests. He argued that because the Moon's gravity was a sixth of Earth's, trees and creatures on the Moon, 'Lunarians', could be six times as tall as on Earth and may erect buildings so large that they would be visible from Earth. He also claimed that the lunar craters were towns erected by the Lunarians and urged astronomers to make a complete census of the circular features so that they could identify new structures built in the future.

It became generally accepted that our Moon hosted life, and several novel methods were proposed to signal our presence or communicate with the life on its surface. One proposal was to construct an immense triangle and squares to represent Pythagoras' theorem. Constructed in pine forests that would be planted in the Siberian tundra, it would be visible to life on the Moon thus revealing our intelligence! In other proposals, large mirrors or giant fires would be used to communicate with the Lunarians.

In 1822 the German astronomer Franz von Gruithuisen dramatically announced that he had discovered a lunar city on the borders of the Sinus Medii, not far from the centre of the disk. He described a collection of dark gigantic ramparts stretching 37 kilometres away, arranged like a work of art. His ramparts later turned out to be rather haphazard lunar ridges called rilles.

Wilhelm Beer and Johann Heinrich von Maedler began their selenography work in 1829, and spent 600 nights over six years to create a one metre sized chart of the Moon. It was an impressive accomplishment, carried out with a small but high quality 10-centimetre telescope that could resolve features on the Moon about ten kilometres across. Despite the arguments to the contrary by most of the earlier selenographers, Beer and Maedler thought the Moon was a lifeless world. It might have been the last word on the subject and the last lunar map, but 30 years later the astronomer Julius Schmidt announced that one of the craters observed by himself, and recorded by Beer and von Maedler, was no longer there. The race to search for changes on the Moon began again, ever larger telescopes were trained on the Moon, and even more maps were drawn.

John Herschel, the son of William Herschel, took a powerful reflecting telescope to Cape Town in 1833. His aim was to observe the southern skies since most of the best telescopes had previously been located in the Northern hemisphere. Soon after, Richard Locke, a reporter at the '*New York Sun*', took a chance at writing a series of spoof stories since Herschel was far away and would not be able to immediately refute them. In his first piece he convincingly wrote that Herschel had modified his telescope with a completely new principle and such magnifying power that he had discovered remarkable forms of life on the Moon. He promised more to come. In the next article he wrote about Herschel's observations of amethyst mountains, sapphire hills with flying unicorns, ape-men with bat-like wings and strange rolling amphibious monsters. Incredibly, not only were the public fooled, so were the media and many scientists. The '*New York Times*' wrote that the new observations "*had created a new era in astronomy and science.*"

By the time that John Herschel heard of the story, it was known across the world and the New York Sun had tripled its readership. He issued a denial but the 'great moon hoax', as it is now called, showed how willing people were to accept the idea that the Moon hosted life.

By the mid-19th century, astronomers suspected that the Moon was dry and airless. The state of lunar knowledge at the time was described in the 1872 essay by the American physician and writer Oliver Holmes '*The poet at the breakfast* table': *"If there were any living creatures there, what odd things they must be. They couldn't have any lungs, nor any hearts. What a pity! Did they ever die? How could they expire if they didn't breathe? Burn up? No air to burn in. Tumble into some of those horrid pits, perhaps, and break all to bits."*

By this time, telescopes had a magnifying power of around six thousand. In a very crude approximation, that brings the surface of the Moon to a distance of 60 kilometres away. With the blurring effect of atmospheric motion, it gives a limit to the size of an object on the Moon that could be resolved of about one kilometre. As the astronomer Richard Proctor stated, that is like viewing Mont Blanc from Geneva, and he writes: *"At this distance the proportions of vast snow-covered hills and rocks are dwarfed almost to nothingness, extensive glaciers are quite imperceptible, and any attempt to recognise the presence of living creatures or of their dwellings is utterly useless."*[96]

Proctor goes on to summarise all the efforts by cartographers of the Moon and their search for features and life analogous to Earth: *"For two centuries and a half, her face has been scanned with the closest possible scrutiny; her features have been portrayed in elaborate maps; many an astronomer has given a large portion of his life to the work of examining craters, plains, mountains, and valleys for the signs of change; but hitherto no certain evidence or rather no evidence save of the most doubtful character has been afforded that the moon is other than a dead and useless waste of extinct volcanoes."*[97]

---

[96] "The Moon: her motions, aspect, scenery and physical condition", R. A. Proctor, 1873, page 242.
[97] Ibid page 259.

## An inhospitable world

At the end of the 19th century and beginning of the 20th century, further observations confirmed that the surface of the Moon is indeed a harsh environment. Astronomers discovered it was a barren world with no atmosphere, a world where liquid water could not exist and where the temperatures varied across an extreme range far beyond the temperatures on Earth.[98]

The first attempt to measure the temperature of the Moon took place in the 17th century. Its light was focused onto a thermometer, but no change in temperature was seen. In 1869, a few years before Proctor wrote his textbook on the Moon, the astronomer Lord Rosse managed to detect the warmth of the full Moon by focussing the Moon's light using a one metre telescope onto a thermopile – an electronic device that is sensitive to temperature. He roughly estimated the daylight temperature on its surface as being above the boiling point of water. By the beginning of the 20th century, the temperature of the bright and dark parts of the Moon had been measured with reasonable accuracy.

In the dark crater depths at the poles of the Moon, temperatures reach minus 247 degrees centigrade – the coldest place ever measured in our solar system. Compared to that, the average night time temperatures on the Moon would be quite a relief at minus 183 degrees centigrade. The average day time temperatures are a steamy hot 106 degrees centigrade. These rather inhospitable extremes are due to the

---

[98] "Water on the Moon, I. Historical Overview", 2011, A. Crotts, Astronomical Review, 6, 4.

fact the Moon has almost no atmosphere, and heat is rapidly radiated away into space. On Earth the difference between day and night temperatures is far smaller thanks to our insulating atmosphere.

The drop in temperature on the Moon when the Sun sets is very rapid. This was first measured during a lunar eclipse, when in a matter of one hour the surface had radiated away most of its heat. In 1949 radio waves were measured coming from the Moon – not alien transmissions, but just because it's a warm rock and anything that has heat will radiate in the infra-red to radio wavelengths. When measured with radio wavelengths, the temperature difference between night and day on the Moon was much smaller, and it took several days after sunset for the coldest temperatures to be reached.

The radio waves were measuring stored heat from many metres below the surface of the Moon, which implied that the very top layer of material on the Moon was acting as an insulator, trapping the heat from below the surface. There were two interesting consequences to this. Whereas a lunar settlement on the surface would suffer drastic temperature changes from night to day, a sub-surface settlement would have relatively steady temperatures that are a little below the freezing point of water. The second consequence was that the top insulating layer of material on the Moon must be rather dusty because it is transparent to radio waves. At the time, it was speculated that the Moon must be covered in a deep layer of volcanic ash since volcanoes were thought to have produced all the lunar craters. This was potentially bad news for landing a space craft on the Moon, and it was the origin of the concerns of the Apollo program – that a landing craft would just sink beneath the surface.

## The vacuum of space

In 1648 the French physicist Blaise Pascal carried a barometer up the Puy de Dome in France. It was quite a feat since the barometers of that time were large and heavy. At the summit, 1460 metres above sea level, he observed that the air pressure was lower – from this experiment Pascal surmised that a vacuum could exist beyond the Earth's atmosphere. And in 1787 instruments were taken to the top of Mont Blanc, measuring the change in temperature and pressure along the way, showing further evidence for a decline with altitude of our atmosphere. However, the most convincing evidence that our atmosphere only extended a few tens of kilometres about the Earth was obtained in 1902. Scientists in France and Germany developed weather balloons that could reach the stratosphere, where the atmospheric air pressure is one thousand times thinner than at sea level.

Early attempts to study the atmosphere of the Moon by Galileo and Huygens failed to find any evidence. Stars were observed occulted by the Moon passing behind the dark edge that was not illuminated by the Sun. The stars would blink out immediately, rather than slowly fading as expected if the Moon had a refracting atmosphere.

In 1864 the astronomer William Huggins observed the spectrum of stars that were occulted by the Moon. By using a prism to observe the colours spread out from blue to red, it was hoped to see the spectrum change as the light from the star passed through the Moon's atmosphere, just like when the Sun sets on Earth, the Sun and sky turn red. No change in the stars' colours was noticed as they set below the horizon of the Moon. No absorption by a Moon atmosphere was observed and no weather phenomenon on the surface of the Moon were observed. In 1892, William Pickering made a series of occultation measurements from which he concluded that any lunar atmosphere must be several thousand times less dense than Earth.

Neither was it really expected that the Moon had an atmosphere, or at least that's what the theorists said. That's because its gravity is one sixth of the Earth, and it would not be strong enough to prevent an atmosphere from escaping into space. The molecules in the air around us move at the speed of sound, that's around 340 metres per second. To escape Earth's gravitational pull, molecules would have to move at 11 kilometres per second, whereas on the Moon the escape velocity is just 2.4 kilometres per second – a speed that would be reached once molecules are warmed by the sunlight.

With such extreme temperatures and no signs of an atmosphere, by the start of the 20$^{th}$ century most astronomers had given up on the idea that life could exist on the surface of the Moon. But a few scientists held out the belief. The last serious advocate of animal life on the Moon was not a science fiction writer, but the respectable American astronomer William Pickering. He was the author of an extensive photographic atlas of the Moon in 1904. Observing the Moon from Jamaica in the 1920's, Pickering was convinced he had seen regular variations on the lunar surface. He published a paper in 1924 arguing that the moving patches could be explained by swarms of insects. Pickering used the analogy of an astronomer on the Moon looking at the plains of North America, seeing the herds of buffalo slowly migrating. But because the patches on the Moon changed slowly, he thought swarms of insects more likely. His ideas were featured in the 1929 silent movie '*Frau im Mond*'.

Even shortly before the Moon landings, the idea that life could exist on the Moon was believed by a few. As late as 1959, the amateur astronomer Axel Firsoff concluded in his book '*The strange world of the Moon*', "*To sum up, there does not seem to be any sufficient reason why plants, even of a highly organised type, should be unable to exist on the Moon, though probably only in isolated oases of life, the highlands being almost entirely barren as they appear to be on Mars.*" I don't know how Firscoff thought that plants could survive in a vacuum, being baked in the Sun at over 100 degrees centigrade.

Even above boiling point, some types of extremophilic microbes can survive. Some species even have optimal growth rates at this temperature. Although the absence of liquid water on the surface of the Moon made the existence of Earth-like microbes impossible, NASA wisely took great precautions upon the return of the first Apollo astronauts from the Moon. After all, who knows what alien microbes might be like!

Despite the clear lack of evidence of an atmosphere or rivers and oceans, there were still many reports of observers seeing 'mists' inside the Mare Imbrium. Features were claimed to disappear out of view. Reports of such low clouds on the Moon spanned from the 1870's to 1950's. In his popular astronomy book '*A guide to the Moon*' (1953), Patrick Moore wrote that although there is not much activity on the surface, here and there we can trace landslips, glows and mists. There were also reports of bright flashes on the surface of the Moon lasting several seconds. These lunar lights were speculated to come from meteors burning up in a very thin lunar atmosphere. But they could have arisen from chance observations of meteorite impacts. However, at this time the lunar craters were assumed to have formed via volcanoes or eruptions of gas, and many had searched for signs of their activity.

### The Moon's craters

In his first reports of viewing the Moon through his telescope in 1610, Galileo wrote in Sidereus Nuncius *"From observations of these spots… I have been led to the opinion and conviction that the surface of the moon is not smooth, uniform, and precisely spherical as a great number of philosophers believe it (and the other heavenly bodies) to be, but is uneven, rough, and full of cavities and prominences, being not unlike the face of the earth, relieved by chains of mountains and deep valleys."* Galileo was clearly countering the ancient Greek view held by Aristotle and others, that the Moon was a perfectly spherical crystalline object.

The circular features on the Moon were later named, in 1791, craters by Schroeter, extending its previous use with volcanoes. The mechanism that created the craters on the Moon was debated for three hundred years.[99] In 1665 Robert

---

[99] "Craters on the Moon from Galileo to Wegener", 1999, C. Koeberl, Earth, Moon and Planets, 85, 209.

Hooke carried out experiments to ascertain their nature. He dropped solid objects into a clay and water mixture, reproducing crater-like features. But he rejected the idea that impacts onto the lunar surface created the circular features, since he had no idea where the impacting bodies could come from. At this time, interplanetary space was thought to be empty. He performed more experiments boiling alabaster and concluded that the craters must have formed from gas erupting from the lunar interior.

Over the next hundred years, volcanic eruptions were commonly thought to be the mechanism behind the formation of the lunar craters. William Herschel even claimed to have witnessed several active volcanoes on the dark part of the crescent Moon in 1787. Gruithuisen advocated the impact origin of craters in 1829, but his claims of seeing inhabited cities on the Moon with cows grazing on lunar meadows did not inspire confidence amongst other astronomers.

The first lengthy study of the impact hypothesis was made by the American geologist Grove Karl Gilbert in 1893. After many experiments he concluded that only impacts could explain the formation of lunar craters, but he published his studies in a journal that was not read by many astronomers. He also commented that the circularity of all lunar craters was a major problem, since some collisions with the lunar surface should take place at an angle and create elliptical craters. In contrast, he rejected the idea that the few known craters on Earth, such as Meteor Crater in Arizona, were made by impacts, rather he concluded they were made by steam explosions.

Steam explosions remained a common explanation of the Moon's craters and the few craters on Earth during the first part of the 20th century. William Pickering describes his lunar observations of ice, snowstorms and seasonal vegetation in his 1903 photographic atlas of the Moon; *'The Moon: A Summary of the Existing Knowledge of our Satellite'*. He explains the craters as the result of steam explosions from beneath a kilometre thick covering of snow. As an eminent astronomer, Pickering was taken seriously, whereas Gilbert was seen as a geologist who dabbled in amateur astronomy. This is illustrated by the well-known story of a local politician who criticised that the U.S. Geological Survey had so little work that one of its most prominent members had nothing better to do than observe the Moon all night long – which led Gilbert to remark that *"clouds and politicians are equal hindrances to serious work."*

An interesting claim was made in 1903 by Nathaniel Shaler, professor of Paleontology and Geology at Harvard University: *"The fall of a bolide of event ten miles in diameter would have been sufficient to destroy organic life on Earth."* However, Shaler was arguing against the impact origin of lunar craters since he continues his text *"…yet life has evidently been continued without interruption since before the Cambrian time."*[100] At that time the periods of mass extinctions on Earth were not known since radiocarbon dating of fossils was not developed until the 1940's.

In 1916 the Estonian astronomer Ernest Öpik came up with the reason why impact craters on the Moon would all be

---

[100] "Shooting the Moon: understanding the history of lunar impact theories", 1998, P.H. Schultz, History of Earth sciences society, 17, 92.

circular. Because of the high velocities at which a body hits the Moon, the impacts have such a high energy that they are similar to explosions, which always produce circular craters. Unfortunately he published his findings in Russian within a little read Estonian journal. In 1921, Alfred Wegener discussed his experiments on impact craters and discovered the formation of central peaks and the larger scale splashes that occur – the crater rays that form from the ejected debris skimming across the lunar surface. But just like his ideas on continental drift, they were ignored for several decades and for a similar reason – the lack of a causal mechanism. Just as no phenomenon was known that could move the continents, no population of objects in space were known to exist that could produce the lunar craters.

At this time some alternative ideas were presented. For example the astronomer Donald Beard in 1925 wrote: *"The five ramparts of Copernicus could not have been formed by any other process than the secular growth of corals and their successive sinking's beneath the ancient Imbrian sea."*[101]

It was around this time when a mining engineer found iron meteorite fragments around Meteor Crater in Arizona. But after a long time searching for the potentially valuable metallic deposits under the crater floor nothing was found. This was taken as evidence amongst the geologists that an impact from space had nothing to do with the creation of Meteor Crater. But astronomers familiar with the results of Öpik knew that a rather small body of about 50 metres was needed to cause a crater of one kilometre in diameter, the size

---

[101] "Coral origin of the lunar craters", 1925, D. Beard, Popular Astronomy, Vol. 33, 1925, p.75.

of the Arizona crater, and it would likely explode into small fragments.

The American planetary scientist Ralph Baldwin added a new piece of evidence to the debate in 1949 in his book '*The face of the Moon*'. It may seem like the lunar craters and World War II have nothing in common, but Baldwin measured the size and depth of bomb craters and found that the distribution of the sizes and depths was exactly the same as had been measured for the Moon using shadows to determine the depths.

Even up to the date of the Moon landings, the astronomical community was divided and the geologists favoured the volcano or steam hypotheses. It was only after impact melted rock samples from the craters were returned to Earth by the Apollo program that the evidence of impacts became clear. Despite this, the geologists still refused to accept that such events had occurred on Earth. In 1980 the physicist Luis Alvarez published the idea that the extinction of the dinosaurs, and a large fraction of all existing species at the time, was due to the impact of a large asteroid or comet 65 million years ago. This was met with disbelief and ridicule from the geology community. But it was the debate that followed this suggestion, which, over the past 20 years, finally led to a more general realisation that impact cratering is an important process on Earth as well, and not only on the other planetary bodies of the solar system.

## 11. Lord of the Tides

Isaac Newton was one of the first true 'theorists' who transformed astronomy by turning it into a science that could make predictions. Some have argued that modern science began with Isaac Newton, although the Dutch physicist and astronomer Christiaan Huygens would at least come a close second. Like all scientists, Newton was building upon the work of others before, and the path to his greatest achievements began by trying to understand the motion of the Moon.

### Apples and cannon balls

In his 1873 astronomy text book, Richard Proctor describes the steps that Isaac Newton took to discover the laws of gravity using the motion of the Moon. He begins with the sentence: *"In the whole history of the researches by which men have endeavoured to master the secrets of nature, no chapter is more encouraging than that which relates to the interpretation of the lunar motions."*[102]

Isaac Newton was a student at Cambridge University, but in the year 1665, when he was 23, the university was closed because of the great bubonic plague. He spent a year at his parents' home thinking, something that scientists these days have less time to do thanks to an overload of bureaucracy, committee meetings and writing funding applications!

---

[102] "The Moon: her motions, aspect, scenery and physical condition", R. A. Proctor, 1873, page 137.

It was during this time that Newton came up with his three laws of motion and the basic workings of the gravitational force. His laws of motion were an extension of similar ideas by Galileo and Descartes. His theory of gravitation was influenced by watching an apple fall – a true story recorded by himself and several others at the time. He wondered if the same force that pulls the apple to the ground could extend to the Moon. But while the apple falls straight down because it has no initial motion, Newton's brilliant insight was to wonder if that same force caused the Moon to constantly fall towards Earth, but its motion around and away from the Earth exactly balanced the falling.

His thought experiment, which he sketched in his famous work '*Principia*', is now called 'Newton's cannon'. It shows a cannonball fired from the top of a mountain at higher and higher speeds horizontal to Earth's surface. As the speed increases the cannonball lands further away from the mountain, first falling in front of the mountain, then reaching a small way around the planet, then a bit further, until at sufficient speed the cannonball orbits the Earth. In the absence of friction from the air, this is what would happen if the cannonball were shot at a horizontal speed of eight kilometres per second.

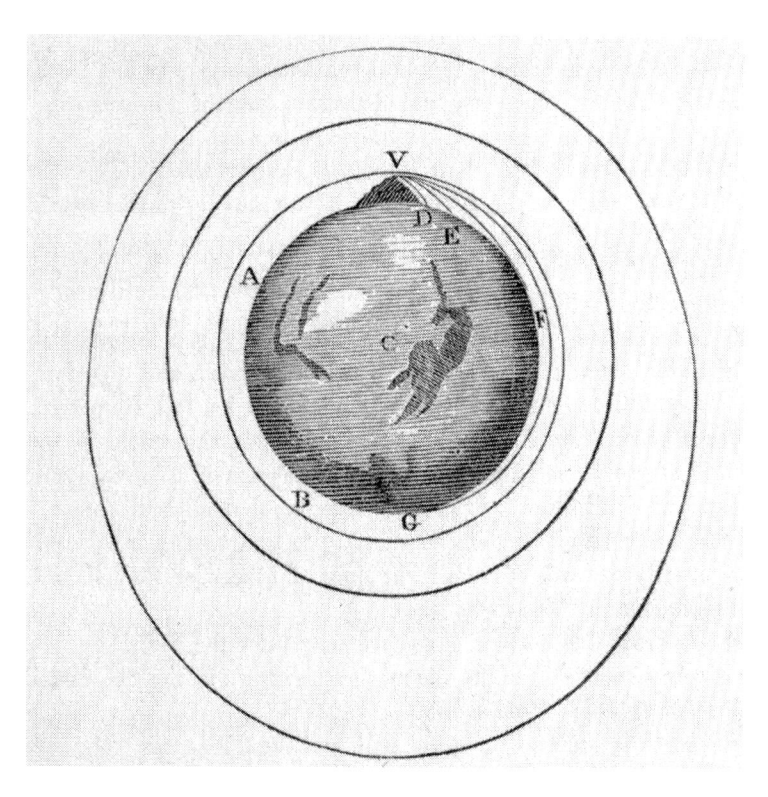

*Newton's cannonball from 'Principia'*

Newton extended the idea of the cannonball to the motion of the Moon. His reasoning would have gone something like this: At its orbital speed of one kilometre per second, if it were not for the force of gravity, the Moon would move in a straight line a distance of one kilometre in one second. And using simple geometry Newton would have calculated that during that one second the Moon would have moved about 1.4 millimetres away from the Earth. But to stay in a circular orbit around the Earth, the gravitational force of the Earth must cause the Moon to fall back to Earth by exactly the same amount, 1.4 millimetres, each second.

Newton proposed that both the apple and the Moon are falling because of the attractive force of the Earth, and that force decreases in intensity just like a beam of light. Although Kepler also knew that the intensity of light decreases as the inverse square of the distance, he thought that the force that maintained the motion of the Moon fell in proportion to the distance away. Even before Newton, in his 1645 book *'Astronomia Philolaica'*, the French astronomer Ismaël Bullialdus correctly argued that the power by which the Sun holds the planets weakens with the square of the distance away. The English philosopher Robert Hooke had also written a letter to Newton explaining his ideas of a universal law of gravitation that decreased with distance in the same way.

It seemed that many of the theorists of the day had similar ideas in mind. However, Newton turned the ideas into equations and showed that the inverse square force was the only sensible force law that naturally gave rise to the elliptical orbits of the Moon and planets. In fact, the only other force law that produces elliptical motions is a force that increases with distance, and that would make no sense. All other possible force laws do not give rise to perfect elliptical orbits.

But the numbers did not quite work out. Knowing how far the Moon had to fall to Earth in one second, he calculated how far the apple should fall in one second on Earth's surface. Since the Moon was 60 times the Earth radius away, the apple should fall 60 multiplied by 60 times further in one second.

Using the distance the Moon had to fall to Earth in one second to maintain its orbit, he calculated that the apple should fall 4.3 metres in one second on Earth, not the measured 4.9 metres. It doesn't seem like a large error, about thirteen percent, but Newton could not find out where his reasoning was wrong. And that was enough of an error to cause Newton to put his ideas on hold for over a decade.

Newton's value for the distance to the Moon as being 60 Earth radii was close to the correct value. The problem was in the measurement of the size of the Earth. In the mid-16th century the circumference of the Earth was measured to be 34,762 kilometres – unbeknownst to Newton, this value was thirteen percent smaller than the true value.

Interestingly, the value of the size of the Earth seems to have been measured more accurately by Eratosthenes in the 3rd century BC, who got the answer correct to about one percent using the shadow of a gnomon. Perhaps Newton did not trust this measurement. Even more remarkable is the Indian astronomer Aryabhata. In the 7th century AD he quotes the size of the Earth accurate to a fraction of a percent. Unfortunately there are no details on how he performed this measurement, and it was probably unknown to Newton.

In 1684 Newton heard news of a new measurement of the size of the Earth by the French astronomer Jean Picard that was thirteen percent larger than he had originally used! Picard had measured the distance between two points at the same longitude, but differing in latitude by one degree, and multiplied the distance by 360 to determine the circumference of the Earth. When Newton used Picard's value in his calculation that compared the falling motion of the Apple and the Moon, the numbers worked out exactly right. Since the apple and the Moon fell towards Earth at the expected rates, Newton had proved that Earth's gravitational force caused the motion of both.

Newton went on to publish the first version of *'Philosophiae naturalis principia mathematica'*, also known as *'Principia'*, in 1687 and revised extended versions in 1713 and 1726. Despite Newton being a disagreeable person, his scientific insights were remarkable. He not only showed how falling objects on Earth and the lunar and planetary orbits could be explained with his theory of gravity, he did far more. With his orbital theory, he could calculate the distances to the planets, the mass of the Sun and the motions of comets. He showed how the Moon was responsible for the ocean tides and explained why there are two tides a day. And there was even more.

Newton was also the first to tackle the impossibly difficult three body problem – the Earth, Moon and Sun system. Although he understood the basic elliptical orbit of the Moon, he could not explain the peculiar precessing and wobbling shape of the Moon's orbit, which is partly due to the gravitational force from the Sun on the Moon.

This gives rise to the complicated motions I mentioned in the chapter on eclipses that determine when and where a total eclipse of the Sun occurs. Newton's inability to reproduce the details of the orbit of the Moon was one of the main failures in his 'Principia'[103]. It was a problem that took another few

---

[103] "Success and failure in Newton's lunar theory", 2000, A. Cook, Astronomy and Geophysics, 41, 6, 21.

centuries of the greatest mathematicians and astronomers to completely solve. The contest to reproduce the motion of the Moon was tackled by nearly all the leading mathematicians and astronomers of the 19th and 20th centuries, including Laplace, Lagrange, Euler, d'Alembert, Airy, Delaunay, Poincaré and Newcomb, to name a few.

Newton began his investigations wondering if gravity would extend to the Moon. If Earth had no Moon, it might have taken far longer for someone to prove the correct law of gravitation. His proof of concept relied on knowing the distance to the Moon for which we can thank the ancient Greeks. The geometrical laws to the planetary motions that Newton used were thanks to Kepler. The foundations of the laws of motion were thanks to Descartes and Galileo. And for making him think about forces, Newton should have really acknowledged Hooke.

**The ocean tides**

The connection between the phases of the Moon and the ocean tides was probably made by many ancient cultures that lived by one of the great oceans, where the ocean can rise and fall by several metres about twice a day. But they lacked the symbolism or written language to preserve their observations, so there is no record. The aboriginal stories passed down through generations connect the Moon and the tides, but the age of these stories is impossible to ascertain. The early Sumerian, Indian and Egyptian civilisations lived far from the oceans. It would have been difficult to make the connection living by the Mediterranean, where the tides are just centimetres high. That's because the Mediterranean is more like a large lake, connected to the Atlantic Ocean via a narrow entrance just 13 kilometres wide. The Moon has little effect on a body of water a thousand kilometres across, and the narrow entrance to the Atlantic restricts the flow of ocean water.

Around the year 330BC the Greek astronomer and explorer Pytheas of Massalia made a long voyage northwards, leaving the confines of the Mediterranean Sea.[104] He circumnavigated the British Isles and travelled far north until he came to a point where the Sun never set. He wrote a book called '*On the Ocean*', of which there are no known surviving copies, but it was referenced by many other authors. Pytheas noticed that there were two high tides per lunar day, and that the height of the tides depended on the phase of the Moon. The lunar day is the time for the Earth to rotate once relative to the Moon and is 24 hours and 50 minutes. In other words, the Moon rises over the horizon about 50 minutes later each day. This is longer than our 24 hour day because the Moon is orbiting the Earth in the same direction as the Earth spins. Therefore each day you have to wait an additional 50 minutes for the Earth to rotate far enough such that the Moon is in the same position in the sky.

Approximately twice a month, around the time of the new Moon and full Moon, when the Sun, Moon and Earth form a line (a syzygy), the tidal force due to the Sun reinforces that due to the Moon. The tide's range is then at its maximum; this is called the spring tide. It is not named after the season, but

---

[104] "Investigations of tides from the antiquity to Laplace", 2013, V. Deparis et al, chapter 2 of "Tides in astronomy and astrophysics", Lecture notes in Physics 861, Springer.

derives from the meaning 'jump, burst forth, rise', as in a natural spring. The neap tide occurs when the Moon and Sun are at right angles to the Earth, when the Moon is only half illuminated by the Sun and the difference between high and low tide is the smallest.

But what causes this rise and fall of the oceans that takes place twice per day and depends on the position of the Moon and the Sun? It took almost 2000 years before the correct physical explanation was given, and that was one of the great triumphs of the modern scientific era. But before the problem was correctly solved, there were many failed but innovative attempts to explain the origin of the tides. Seleucus followed the teachings of Aristarchus and correctly believed that the Earth rotated. He suggested that the Moon's motion about the Earth disturbed the atmosphere, which falls as a wind upon the Atlantic Ocean creating the tides. It was not until late in the 17th century that it was speculated the space between the Earth and the Moon was a vacuum.

Another proposal was that the tides were caused by a giant whirlpool, the Maelstrom, off the coast of northern Norway. Low tide was just the consequence of the sea water disappearing into the vortex, while high tide occurred when the water reappeared from the vortex. However, the size of such whirlpools was exaggerated into mythical proportions by painters, poets and writers. With a more scientific approach, in the 13th century, the Arabian scientist Zakariya al-Qazwini wrote *'The wonders of Creation'*, in which he claimed that the tides were caused by the Sun and the Moon heating the ocean waters, making them expand rather like a heated gas that flows to occupy a new space. But he could not explain why the Moon and not the Sun played the leading role.

In 1609 Johannes Kepler claimed that the tides were due to an attractive magnetic-like force from the Moon and Sun. His theory was inspired by the book *'De Magnete'* in 1600 by William Gilbert who concluded that the Earth behaved like a giant magnet. However, Kepler's ideas were criticised by Galileo who argued that that the tides were produced by the combined effect of the Earth's rotation around its axis and its orbital motion around the Sun. These motions would set the water on Earth into oscillations, creating a sloshing motion that is observed as tides, but he couldn't explain the fact that there were two tides per day.

The French mathematician Rene Descartes in 1644 promoted an alternative idea rather similar to that of Seleucus. The Moon and the Earth were each surrounded by a large atmospheric pressure vortex. The pressure exerted by the vortex of the Moon on that of the Earth was transmitted down to the Earth's surface, giving rise to the tides. However, the theory of vortices incorrectly predicted a low tide when there was in fact a high tide.

**Why two tides per day?**

Despite some of the greatest minds of the 17th century tackling the problem, no one had a satisfactory explanation for why there are two ocean tides per day – it all seemed very confusing. The correct solution was eventually given by Isaac Newton, who showed that the tides were a consequence of the gravitational force from the Moon and Sun. With his new theory of gravitation, Newton could explain why there are two main tides per lunar day and why the height of the tides depended on the relative positions of the Moon and Sun. He went even further by using the difference in height of the spring and neap tides to make the first estimate of the mass of the Moon.

This is the point at which things become hard to visualise, but let me try to explain the origin of the ocean tides, first by looking at the strongest driver of the tides, the Moon.

The gravitational attraction of the Moon on material inside and on the surface of our planet varies according to how far that point is away from the Moon (as the inverse square of the distance). When the Moon is overhead it actually lifts you off the ground a little. It's a tiny effect, but it's real. This means that when the Moon is overhead your weight is actually a little less, but it's less than a gram and you can't measure that using a normal set of scales, since the scales are also pulled towards the Moon by the same force. At the same time on the opposite side of the Earth the Moon's gravity is a little weaker because it is an extra Earth diameter more distant.

A common misconception is that it is the Moon's gravity that lifts the ocean water upwards towards the Moon. However, the strength of the Moon's gravity on the surface of the Earth is only about one ten millionth as strong as Earth's gravity that holds everything to the surface of our planet. And this would not explain why there is a high tide occurring simultaneously on the opposite side of the Earth.

The tides are due to the variation of the gravity of the Moon across our planet, which distorts the shape of the Earth and oceans into something resembling a rugby ball. The two bulges, one on the side of Earth closest to the Moon and one on the opposite side of the Earth, are aligned in the direction of the Moon. The rotation of our Earth plays no role in generating this distortion of our planet, but as the Earth rotates these bulges stay pointing towards the Moon, and new regions of oceans (and land) rise and fall.

It is nothing as dramatic as a real rugby ball, the ocean bulges are each about half a metre high compared to the 12.7 million metre diameter of the Earth. So the tidally deformed Earth is actually spherical to one part in ten million. But that half a metre of distortion has some remarkable consequences. Those tidal bulges result in the ocean tides, which are due to the motion of the water into and out of the bulges. They lead to a slow change in the length of the day, they cause the Moon to move away from the Earth and they are the reason we only see one side of the Moon.

So what's really taking place?

The Moon and Earth orbit around their common centre of mass – a point that lies about 1000 kilometres beneath the surface of the Earth in the direction of the Moon. If the Earth and Moon were the same mass, they would orbit around a point half way between them, because both would provide the same gravitational attraction on each other. But the Moon is much smaller, so its gravitational pull on the Earth is far less than that of the Earth on the Moon. Both the Moon and the Earth are continuously falling towards their common centre of mass, and the falling motion is balanced by their orbital motion around the centre of mass – in essence, the

Earth and Moon are both behaving like Newton's cannon ball, orbiting the centre of mass.

The force that moves the ocean waters is the gravity of the Moon – both the Earth and everything on its surface are trying to fall towards the Moon. The water is free to move around on Earth's surface, it flows across the surface of the Earth towards the top of both bulges. Because the force of gravity is stronger for the water on the side closest to the Moon it falls faster towards the Moon and rises up the near-side tidal bulge. And because the gravitational attraction is weaker on the far-side, the water falls more slowly towards the Moon than the rest of the Earth. This is the tricky part to visualise – because the far-side water falls more slowly than the Earth underneath, it is left behind and falls behind, moving up the far-side bulge. Once the ground has rotated away from the bulges the water flows back to its equilibrium position. So if you are watching the ocean tides over the course of a day, you and the ground beneath you will pass through both tidal bulges each lunar day, and the ocean's water will rise and fall around you twice a day.

Another way to look at the motion of the ocean water is to imagine yourself and two friends falling through space directly towards the Moon. One of your friends is in front of you, closer to the Moon, and one is behind you, more distant from the Moon. Because the friend in front of you feels a slightly stronger gravitational pull from the Moon, they fall slightly faster towards the Moon than you, whilst the friend behind you feels a slightly weaker gravitational pull and falls slightly slower than you. What do you see? You see both friends slowly moving away from you, as if you were all being pulled and stretched apart by some mysterious force. That's the tidal force, and it is the result of gravity becoming weaker with distance.[105]

The tidal force is due to the difference between the gravitational pull of the Moon on opposite sides of the Earth. That's why smaller bodies of water do not have large tides – the difference between the Moon's gravity across a swimming pool or a lake is far too small to cause the water to move. The ocean bulges are tiny in height but vast in extent, averaging about half a metre high but spread over several thousand kilometres. However, unlike the ground beneath you the water can flow across the surface, piling up in the tidal bulges.

---

[105] You can also interpret the tides by including the centrifugal force. The average pull toward the Moon is balanced by the Earth's orbital motion around the common centre of mass. This is the centrifugal force, and that force is pointing away from the Moon. If you subtract the Moon's gravitational force acting on the centre of the Earth from the centrifugal force you are left with the force called the tidal force. It causes the Earth to be stretched in the direction pointing to the Moon and compressed in the perpendicular direction.

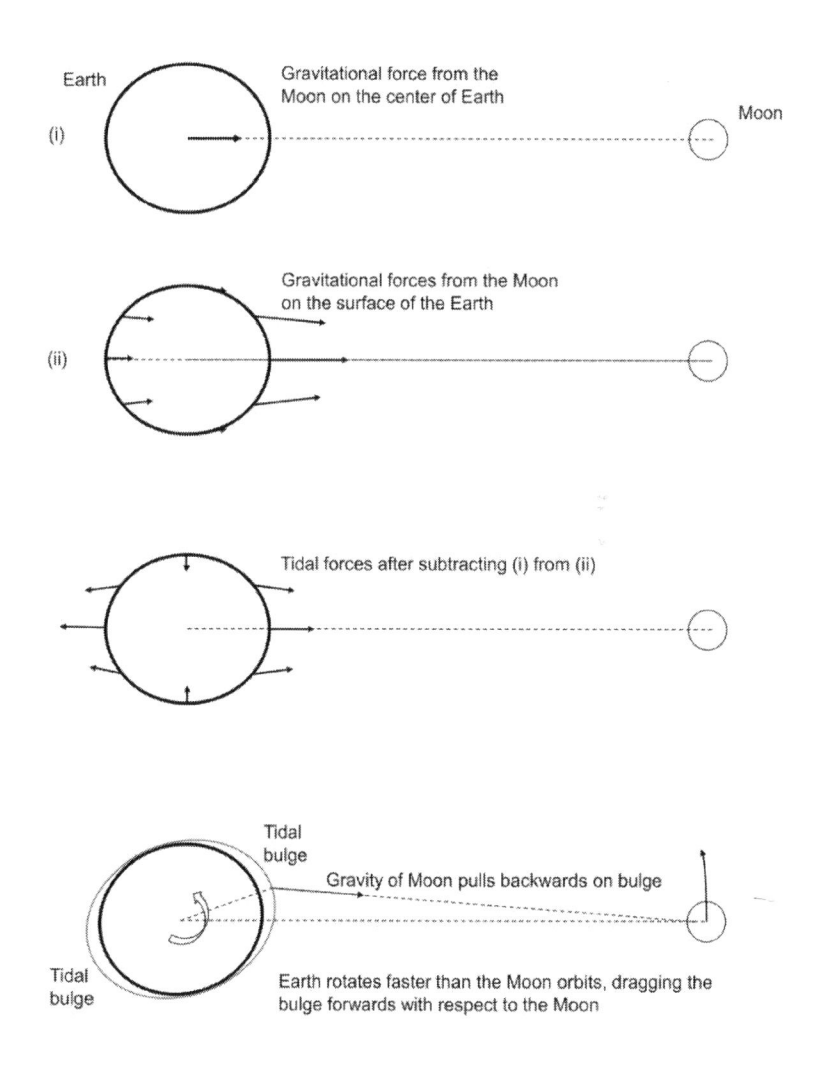

**Earth** (i) — Gravitational force from the Moon on the center of Earth — **Moon**

(ii) Gravitational forces from the Moon on the surface of the Earth

Tidal forces after subtracting (i) from (ii)

Tidal bulge

Gravity of Moon pulls backwards on bulge

Tidal bulge

Earth rotates faster than the Moon orbits, dragging the bulge forwards with respect to the Moon

*The tidal forces on the Earth from the Moon*

Phew. If it takes me more than one paragraph to explain something, you can tell it's quite a tricky thing to explain. Well done for staying with me and not skipping to the next part!

The Sun causes a second set of tidal bulges that point in the direction of the Sun, but the strength of those tides is less than half of those due to the Moon. Even though the Sun is much more massive than the Moon, it is much further away, so the difference between the Sun's gravity on opposite sides of the Earth is not as large as the difference of the Moon's gravity on opposite sides of the Earth. The strongest ocean tides occur during a syzygy, when the Moon and Sun reinforce each other's tidal effects.

Newton's study of the tides was greatly simplified because he treated the water as an equilibrium layer on top of a rigid spherical planet. A more complete mathematical analysis only came in 1799 thanks to the French mathematician Pierre de Laplace. Laplace did this in a very elegant way by showing how the height of the tides at a particular time, latitude and longitude was the sum of several periodically varying components. These include the orbital time of the Moon, the time it takes for Earth to rotate, the time for the Moon to complete an elliptical orbit of the Earth, the time over which the inclination of the Moon's orbit relative to Earth's equator changes, the time it takes the Earth to make its elliptical orbit about the Sun and the variation of the tilt of Earth's spin with respect to its orbital plane about the Sun.

All of this is greatly complicated by the fact that the ocean does not cover the entire Earth but is divided into zones of different depths by the continents. This means that the ocean tides can vary enormously from place to place – such as the Bay of Fundy in Canada where the tidal range can reach over 16 metres due to the geometry of the ocean bay. In some places, such as the Gulf of Mexico, this results in just one tide per day – called diurnal tides. In other places, such as the West coast of America, the two daily tides have unequal heights – called mixed semidiurnal tides. However, the most common tides are the semidiurnal type, with two similar sized tides each day.

Laplace showed that the energy in the ocean tides is dominated by three main periodic components. At many locations most of the energy in the movement of the ocean water is found at three main semidiurnal frequencies: 1.93 cycles per day (due to the Moon's orbital time), 2.00 cycles per day (due to the Earth's rotational time), and 1.90 cycles per day (due to the eccentricity of the Moon's orbit). In some locations, three other diurnal frequencies, due to the asymmetry introduced by the tilt of Earth's axis, play an important role. Laplace's findings would be at the heart of all future tide-prediction methods. Ultimately, it would lead to one of the most elegant mechanical computing machines ever invented.

Laplace's technique for developing practical tide prediction was first taken up 80 years later in Britain by William Thomson (who became Lord Kelvin in 1892). Thomson took Laplace's ideas one step further and came up with a method (called harmonic analysis) for determining how much energy there was at a specific frequency by simply measuring the change in tidal height with time at a specific location. Those energies vary from place to place because of the way oceans, bays and landmasses affect the ocean currents. But the brilliance of Thomson's method was that it required no understanding of the complex topic of gravity and hydrodynamics. You just needed a few measurements at each location to fix the model.

In the early 1870s, Thomson designed an ingenious mechanical computer to automate the prediction process and calculate the height and time of the tides at a given location. It had dozens of gears and pulleys over which ran a wire that was connected to a pen touching a moving roll of paper. Each of the astronomical frequencies was represented by an individual gear that rotated with a speed corresponding to one of the constituent timescales. A pin-and-yoke arrangement transformed the gear's rotation into an up-and-down motion that pulled on the wire, thus providing that component's contribution to the tide curve being drawn on the moving paper.

Thomson's first machine, built in London in 1872, simultaneously calculated the contributions of the 10 most important tidal constituents. It was a large, finely crafted brass apparatus which later became known as 'Kelvin's tide machine'.

Twenty-five years later, Edward Roberts, who had worked out the gear ratios for Kelvin's original device, designed a 40-constituent tide-predicting machine for the Bidston Observatory's Tidal Institute in Liverpool. With such machines, the heights and times of the tides were predicted a year in advance for all major ports and harbours around the world. This knowledge was important for the movement of ships in and out of the ports, and it played a vital role in the Second World War.[106]

## Tides and D-Day

In the spring of 1944 the allied forces of World War II were planning the invasion of Nazi-occupied France. The ocean tides posed a big problem. Along the entire French coast of the English Channel, the vertical range from low tide to the next high tide exceeded six meters. At low tide, those large tidal ranges exposed long stretches of beach that soldiers would have to cross under heavy fire. German Field Marshal Erwin Rommel was convinced that the invasion would occur at high tide so that the invading troops would minimise their exposure time on the beaches. He ordered the construction of thousands of underwater obstacles that would be just below the water at high tide. These were concrete and steel barriers or large wooden spikes designed to rip open the bottoms of any incoming landing boats.

Fortunately, the allied aerial reconnaissance spotted the obstacles at low tide and recognised their purpose. That led to significant changes in the invasion plan and made its success even more dependent on accurate tide predictions. It was decided that the initial beach landings should take place soon after low tide so that demolition teams could blow up enough obstacles to open corridors through which the landing craft could navigate to the beach. The tide also had to be rising, because the landing craft had to unload troops and then depart without danger of being stranded by a receding

---

[106] "The tide predictions for D-Day", 2011, B. Parker, Physics Today, 64, 9, 35.

tide. There were also non-tidal constraints. For secrecy, the allied forces had to cross the English Channel in darkness. But naval artillery needed about an hour of daylight to bombard the coast before the landings. Therefore, low tide had to coincide with first light, with the landings to begin one hour later.

A six metre tidal range meant that water would rise at a rate of at least one metre per hour. The times of low water and the speed of the tidal rise had to be known rather precisely, or there might not be enough time for the demolition teams to blow up a sufficient number of beach obstacles. Also, the times of the low tide were up to an hour different at each of the five landing beaches, which were separated by about 100 kilometres. Therefore, predictions of the tides were needed at all five locations.

The mechanical tide predicting machines were put into action to make predictions for these locations, but they needed to be initialised with known values of the tides, which had to be measured at each location secretly in advance. It was all a lot of work, but it was vital to get it right. And the mechanical machines were slow, but thanks to the brilliant work of Laplace and Lord Kelvin, they were accurate. The 5th, 6th and 7th of June 1944 were all possible dates when all the factors came together. Bad weather on the 5th led to D-Day taking place on 6th June. The plan worked. The obstacles were dismantled in the early morning hours, and the invasion proceeded with low casualties because the defending forces thought their defences would halt any high tide invasion.

## A deformable Earth

Newton and Laplace had assumed that the solid Earth was a rigid sphere – it was just the water that was responding to the gravitation of the Moon and Sun. At around the same time as tidal friction was being explored, there was an ongoing debate about the internal structure of the Earth. Did the interior of the Earth behave as a rigid solid or a deformable liquid, and if so, is the entire Earth deformed by the tides?

Thompson thought the Earth as a whole must also be distorted by the gravitational forces from the Moon and Sun. His arguments were based on the idea that the Earth was still cooling down from its formation, and its internal structure was a hot liquid magma. One of Thompson's students was George Darwin, son of the evolutionary biologist Charles Darwin. George Darwin followed up on the suggestion of Thompson, who had predicted that the distortion of Earth's interior could be measured. The reason being that if the Earth as a whole deformed, the amplitude of the ocean tides would be lower than expected if the Earth's interior was rigid.

In 1882 George Darwin published his analysis of three decades of tidal observations from ports in England, France and India. He was able to find the ratio of the height of the ocean tide on a deformable Earth to that on a rigid Earth, proving the existence of a global deformation of the Earth caused by the Moon. Darwin went on to study the effects of the Earth's tidal bulge and the friction occurring within Earth's interior as the entire planet is constantly deforming.

If the Earth itself was being deformed by the Moon and not just the oceans, then the crust of the Earth should tilt as it passes through the tidal bulge, and the land beneath us rises and falls. The German astronomer Ernst von Rebeur-Paschwitz spent many years refining the design of a horizontal pendulum so that he could measure this tidal tilt of the land. Horizontal pendulums had been developed in the early 19th century as a tool for seismography and to measure the strength of earthquakes. And between the 17th and early 20th centuries pendulums were important tools for science for measuring time and gravity, as well as providing direct evidence for the rotation of the Earth. In 1851 the French physicist Leon Foucault suspended a 28 kilogram brass pendulum on a 67 metre long wire from the dome of the Pantheon in Paris. The weight of the pendulum enabled it to swing for many hours, and over time the plane in which the pendulum swung slowly rotated. As Descartes and Newton stated, motion in a straight line will continue in a straight line forever unless some external force is acting. Likewise, a pendulum always swings in the same plane relative to the universe as a whole. The pendulum does not know that the Earth is rotating beneath it, and the rotation of the swing of the pendulum occurs because it is the observer standing on the spinning Earth that is rotating around the fixed swing of the pendulum.

The same principles were used to measure the tidal tilt of the Earth. If the Earth beneath the pendulum tilts because of the tidal bulge of the Earth, then the pendulum will be observed to swing in a different plane. The size of the effect is tiny. The solid Earth bulge due to the Moon is only about 10 centimetres spread across a horizontal direction of thousands of kilometres.[107] In terms of an angle, that is a tilt of the Earth's crust of about one millionth of a degree. In 1891 at Potsdam observatory, von Rebeur-Paschwitz observed the tiny tilt of the pendulum with a period and amplitude that corresponded to the predicted distortion of the Earth by the Moon.

Further confirmation of the deformation of the Earth came in 1914, also at the Potsdam observatory, by the Austrian geophysicist and astronomer Wilhelm Schweydar, who measured the change in shape of the Earth with a gravimeter – that is, the change in the strength of gravity as the device moves 10 centimetres further from the centre of the Earth!

---

[107] The height of the ocean bulges is larger because the water can flow across the surface of the Earth and pile up to higher depths.

## 12. Fatal Attraction

After three years of reading theoretical physics and astrophysics at university, I was faced with my final exam paper. A significant fraction of the marks came from answering one of three posed questions. I don't remember two of the questions, but I will never forget the question I chose to answer: *"Estimate the height of the ocean tides, the lengthening of the day and the rate at which the Moon moves away from Earth."* Luckily, at the time I was already deeply fascinated by the Moon and was fortunate to have studied Newton's Principia and Laplace's work.

After a sweaty hour of playing with the equations I had come up with a formula to describe the shape of the tidally deformed Earth. I was relieved to finally calculate a tidal bulge height of about half a metre. I could also recall how the tides affect the long term behaviour of the Earth and Moon. By using the fact that energy and momentum are conserved I could estimate how the length of Earth's day varies and how the orbit of the Moon changes in the future. Even though I was happy to pass this exam, it was based upon ideas and calculations already made hundreds of years previously, beginning with Immanuel Kant.

### The days are getting longer

In 1754 the philosopher Immanuel Kant published an article in the *'Koenigsberger Nachrichten'*, the weekly magazine of his home town.[108] It was an answer to the yearly prize question of the Prussian academy of sciences. Kant correctly speculated that the friction caused by the tidal motion of the ocean waters across the surface of the Earth could slow down the spin of the Earth. As a consequence, the length of the day should slowly increase. Kant went on to speculate that the rotation of the Earth should decrease until it always turns the same side towards the Moon, that is, until the length of the day increases to match the orbital timescale of the Moon, around 1000 hours.

The idea of Kant was impressive, but no one really followed it up, perhaps because he did not publish this in a scientific journal.

Let me explain in a bit more detail why our Earth's rotation is slowing down. It takes time for the ocean and solid Earth tidal bulges to return to equilibrium. Because the Earth is rotating relatively quickly, the tidal bulges are dragged forwards so that they do not point directly at the Moon. The Moon's gravity pulls on this bulge, trying to drag it backwards, slowing down Earth's rotation and lengthening our day. This also causes friction between the ocean water and the land at the bottom of the ocean, which generates heat that is radiated away into space. There is also friction within Earth's interior. All of this friction amounts to about three terawatts of energy – that's a thousand times the entire energy consumption of the world!

---

[108] "Ob die Erde in ihrer Umdrehung um die Achse einige Veraenderung erlitten habe und woraus man sich ihrer versichern koenne", 1754, Koenigsberger Nachrichten

At the same time trillions of tons of water are piled up at the bulges, and all of that mass exerts a gravitational force on the Moon. This small additional gravitational pull on the Moon causes its orbital distance from Earth to increase: it moves away from the Earth. In physics speak, energy and angular momentum from the rotation of the Earth are transferred to the Moon.

A century after Kant, the German physicist Robert Mayer and the American oceanographer William Ferrel independently took interest in the effects of the Moon on the length of our day. In addition to calculating the rate at which the Earth's spin should slow down, they pointed out that this effect could be observed in the apparent positions of the celestial bodies. They made the first calculations of the lengthening of the day that occurs because of the Moon. It's a very difficult calculation and depends on many complex factors – the actual measured value of the lengthening of Earth's day is about one second in 50,000 years.

This all sounds rather spectacular, but is there any evidence for the changing length of day on Earth or the increase in distance to our Moon?

Until the 1960's the distance to the Moon was measured using trigonometry applied to lunar eclipses or parallax by measuring the angle between the Moon and distant stars from multiple locations. These techniques were not very accurate, and precise measurements of the distance to the Moon only became possible thanks to the Apollo program.

The Apollo 11, 14 and 15 missions left parabolic reflecting mirrors behind to allow accurate measurements of the distance to the Moon. Pulses of laser light are sent from Earth towards the Moon, and the reflected light is detected using a large telescope. The time that it takes for the light to travel to the Moon and back gives the distance to the Moon. This is not a simple measurement. Even the light of a laser beam spreads out in space, and at the distance of the Moon only one in a billion photons hits the mirrors. The reflected photons also spread out in space such that less than one billionth of those are detected by telescopes back on Earth. However, these can be detected, and we now know that the Moon is moving away from Earth at a rate of about 3.8 centimetres per year – as tidal theory predicted.

What about the lengthening of the day? One second in 50,000 years doesn't sound like much, but it adds up. Imagine a clock that was set to the length of the day defined by the Earth's rotation speed 2000 years ago. If that clock was still running it would be several hours ahead. Thanks to the ancient astronomers' dedication at recording eclipses, we do have access to a clock that was set thousands of years ago.

Because solar eclipses always occur during a New Moon, the time between eclipses is always an integer number of lunar months. In 1695 the astronomer Edmund Halley was comparing the dates and times between eclipses that had occurred over a thousand years ago. The oldest dated solar eclipse occurred in China in the year 708BC and the oldest reliably dated lunar eclipse in Babylon in the year 665 BC. The times of day that these eclipses occurred were recorded to within a fraction of an hour. Halley noticed that there was a systematic offset of several hours between the measured and expected times of the eclipses.

With this information Halley concluded that the Moon's motion is accelerating. That was a plausible but incorrect interpretation. It was not the motion of the Moon that was changing, but the length of the day. If the length of the day had been constant since this time, then the eclipses would have occurred significantly later in the day from when they were observed. From this information the change in the length of day over the past 2000 years can be calculated, and that turns out to be 1.8 milliseconds per century.

Nowadays the rotation rate of the Earth can also be measured by observing when far away astronomical objects reach the same position in the night's sky. Since 1960 astronomers have been accurately measuring the positions of pulsars – cores of dead stars that emit clockwork-like pulses of radio waves. From these observations astronomers have directly measured that the rotation speed of the Earth is slowing down such that the length of the day is increasing by 1.7 milliseconds per century.

If our understanding of tidal theory is correct then the length of the day must have been much shorter millions of years ago. It is difficult to predict exactly how much the rotation rate of the Earth changed in the past or will change in the future, because most of the energy loss is due to the friction of the water in shallow oceans less than one hundred metres deep. And where those shallow oceans occur depends on the past and future configurations of the continental plates which are haphazardly drifting around on the surface of the Earth.

Indeed, there is evidence from the fossil record that the length of the day was significantly shorter hundreds of millions of years ago. Corals, like tree trunks, also bear records of growth periods since they add microscopically thin layers of calcium carbonate each day. The size of these layers varies according to the seasons, since they reveal when the corals were growing rapidly and when they weren't. The lines can help us differentiate between the busy growing seasons from year to year and even from day to day. You can also see how they stack up into monthly deposits linked to the lunar cycle, since some corals build new skeleton partitions each lunar month. And those days per year are different depending on when the corals lived. Fossilised corals from the Silurian Period, 444-419 million years ago, show 420 little lines between seasonality bands, indicating that a year during that period was 420 days long. More recent corals from the Devonian Period, a few million years later, show that the earth's spin had slowed down to 410 days per year. Today, corals have 365 growth rings a year. Putting all this together means that the day was about 20 hours long 400 million years ago, confirming the prediction that tidal friction has constantly slowed down the spin of our planet.

### Why we only see one side of the Moon

Now that we have all this knowledge of the tides and tidal bulges, we can use it to understand why we only see one side of the Moon.

The so called dark side of the Moon is a terrible name. Since it is not dark – it receives as much sunlight as the near side of the Moon – it should just be called the far side of the

Moon. We only see one side of the Moon because the Moon spins very slowly – exactly once in one orbit about the Earth. This is no coincidence and the reasons were worked out by those scientists in the 19th century who had calculated that tidal friction causes the Moon to slowly drift away from Earth.

In the far future the rotation of the Earth will slow down so much that it rotates exactly once per lunar orbit. This is called 'tidal locking'. It has already happened to our Moon, and that's why we only see the near side of the Moon. If the Moon was once spinning faster, the same tidal effects of the Earth on the Moon would have caused the Moon's rotation to slow down, thanks to the gravity of Earth pulling the Moon's bulge backwards. This slowing down of the Moon's rotation would have continued until it became tidally locked such that one side of the Moon permanently faces the Earth. Once tidally locked, the Earth still causes tidal bulges on the Moon but these point directly at the Earth. If the Moon was once rotating more slowly than its orbital time, the tides would have caused its rotation to speed up until it became tidally locked.

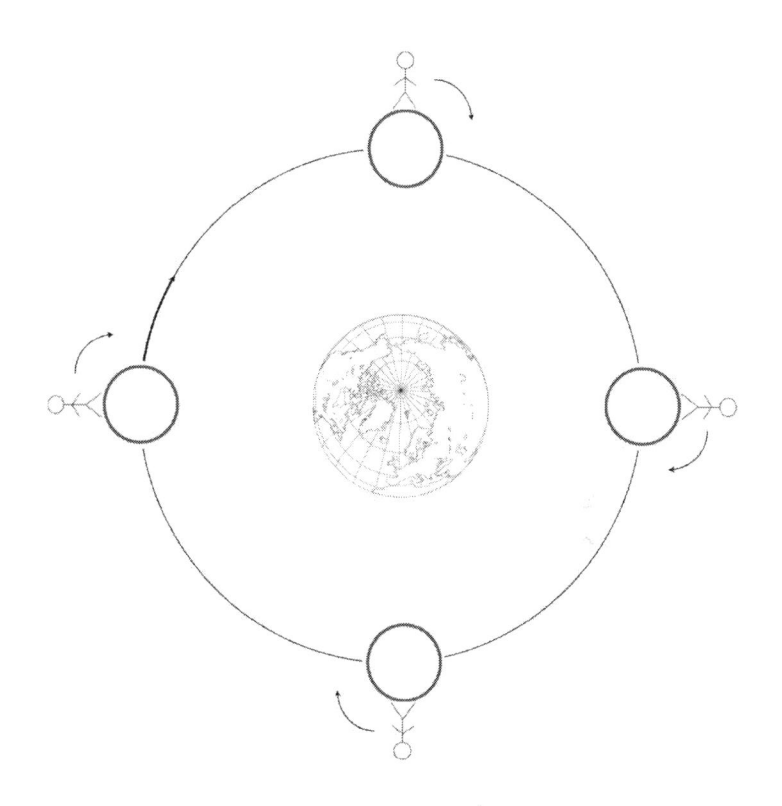

*The tidally locked Moon*

This incredible synchronisation of orbital motion and rotation occurs in other solar system objects. For example, Pluto and its Moon Charon are fully tidally locked so that their orbital period and both of their rotation periods are all 6.4 days. And the recently discovered exoplanets in the habitable zone of red dwarf stars are all thought to be tidally locked, with one side of their worlds facing the star and the other side in permanent darkness.

Another interesting effect occurs if a planet's moon is orbiting in the opposite direction to the rotation of its planet. In this case, the physics is reversed, and the tidal bulge lags behind the direction of motion. This causes the rotation speed of the planet to increase and the moon to drift towards the planet until it eventually comes so close that it either crashes onto the surface of the planet or is torn into fragments. This is actually happening to one of the moons of Neptune. Triton orbits Neptune in the opposite direction to Neptune's spin. It is already tidally locked and is slowly moving closer towards Neptune's surface.

**Gravity and habitability**

The gravitational squeezing of a moon by its planet grows in strength according to the mass of the planet. Saturn is a hundred times the mass of the Earth, and Jupiter is over three hundred times as massive. The gravitational squeezing of their moons is intense – the constant squeezing and stretching of a moon generates heat. It's similar to the reason that a tennis ball or squash ball becomes hot after it has been hit many times. Even if the moons are tidally locked, their slightly eccentric orbits result in a varying gravitational tidal field.

This gravitational heating can be so large that it causes the inside of the moon to be molten hot. The beautiful moons of Jupiter and Saturn, Europa and Enceladus, are known to have warm oceans of water beneath their cold icy surfaces. It is thought the oceans are kept warm because they rest on a rocky mantle that is warmed by the gravitational squeezing by the giant planets they orbit. These moons of Jupiter and

Saturn are prime candidates in the search for alien life in our solar system that may have evolved independently from Earth.

While gravitational tides can make some moons habitable by warming their interiors, they can be so intense that they make the surface of a rocky moon uninhabitable. This has happened to Jupiter's moon Io. It is Jupiter's closest moon, about the size of Earth's Moon, and orbiting above Jupiter's cloud tops at about the same distance from Jupiter that our Moon orbits the Earth. But Jupiter is far more massive, and its gravitational squeezing has caused the surface of Io to be covered with lava flows and over 400 active volcanoes. Some of the volcanoes on Io have been observed to produce plumes of sulphur gases that reach 500 kilometres above its surface.

Director James Cameron could have used some advice on tides from astronomers for his movie Avatar. Much of the action takes place on Pandora, a moon of a hypothetical gas giant planet in the Alpha-Centauri system. Pandora is covered with life and lush vegetation, but at its close proximity to the giant planet its surface would be more like that of Io, a desolate, inhospitable world covered by lava flows and active volcanoes!

### The effects of the Moon on weather

Most of the emerging civilisations paid great attention to the night's sky in search for messages from the gods. The largest surviving record are the thousands of preserved omens and myths from the Babylonians. They tried to find relations between the eclipses and position or appearance of the Moon with storms, wind, rain, drought, earthquakes, volcanoes and pretty much all metrological and Earth-based phenomena. The Zuni Native American peoples thought a red Moon brought water. Seventeenth-century English farmers believed in a 'dripping Moon', which supplied rain depending on whether its crescent was tilted up or down.

'Harvest Moon' and 'Hunter's Moon' are traditional terms for the full Moons occurring during late summer and in the autumn. The 'Harvest Moon' is the full Moon closest to the autumnal equinox. The 'Hunter's Moon' is the full Moon following it. The names were used at least as early as the 18th century. The *Oxford English Dictionary* entry for 'Harvest Moon' cites a 1706 reference, and for 'Hunter's Moon' a 1710 edition of *The British Apollo*, where the term is attributed to 'the country people'.

All full Moons rise around the time of sunset. Because the Moon appears to move eastward faster than the Sun, it rises later each day – on average by about 50 minutes. During the month of March the time between the rising Moon on successive days is over 60 minutes. The Harvest Moon and Hunter's Moon are unique because the time difference between moonrises on successive evenings is much shorter than average. Towards the end of summer in the Northern hemisphere, the Moon rises less than 40 minutes later from one night to the next. That's because of the Moon's orbit around our tilted Earth. At this time there is a shorter period of darkness between sunset and moonrise for several days after the full Moon, thus lengthening the time in the evening when there is enough light for the gathering of the harvests.

The July full Moon has been called the 'Thunder Moon' because of its ancient association with thunderstorms. Given thousands of years of folklore relating the Moon and the weather, is there any evidence for a connection between the two? Well, it turns out that there is.

In 2010 researchers analysed the flow rates of 11,000 streams in America that had measurements dating back to 1900. When they calculated the Moon's phase for each measurement they found a slight increase in water flow around the half Moon, the midway point between the full and new Moons.[109] To show a link to rainfall, the researchers turned to the U.S. Historical Climatology Network, a database with daily precipitation information for more than 1200 stations from as early as 1895. True to farmers' wisdom, precipitation tended to rise a few days before the half Moon! Although it is a small effect, the difference in rainfall and stream levels was about one to two percent higher during this period of the lunar phase. How can this possibly be?

Based on older research, scientists had suggested that the Moon's orbit could distort the Earth's magnetosphere, a region of charged particles surrounding the Earth's protective magnetic field. This might allow more particles from space into the atmosphere, where they could trigger rain when they collide with clouds. But what is actually happening is that when the Moon is overhead it suppresses the amount of rainfall.

---

[109] "Lunar tidal influence on inland river streamflow across the conterminous United States", 2010, R.S. Cerveny et al., Geophysical Research Letters, 37, 22.

New research published in 2016 showed that the small variation in rainfall is due to the gravitational tidal force of the Moon.[110] It was found that the air pressure on the surface of the Earth correlated with the position of the Moon. When the Moon is overhead, its gravity causes the Earth's atmosphere to bulge toward it, so the pressure or weight of the atmosphere on that part of the planet increases. Higher pressure increases the temperature of air parcels below. Since warmer air can hold more moisture, this decreases the rate of rainfall at these times. Using 15 years of data collected by NASA and Japan's Aerospace Exploration Agency, it was found that the rainfall is indeed slightly lighter when the Moon is high in the sky. The change is only about one percent of the total rainfall variation, but it is enough to explain the previous findings on stream flow and rainfall rates.

The Moon also affects the global temperatures on Earth – although it is a very small amount. Satellite measurements of the temperature of the atmosphere show that the poles are half a degree Celsius warmer during a full Moon than during a new Moon. These small temperature changes have a slight but measurable effect on the weather. The temperature increase is due to the infra-red radiation from the Sun reflected by the Moon toward Earth together with the infrared radiation from the Moon itself – the heat that is stored in its surface layers. Although infrared radiation from the Moon is only one-100,000th as intense as the infrared radiation arriving directly from the Sun, much of this radiation is

---

[110] "Rainfall variations induced by the lunar gravitational atmospheric tide and their implications for the relationship between tropical rainfall and humidity", 2016, T. Kohyama and J.M. Wallace, Geophysical Research Letters, 32, 918.

absorbed in the lower atmosphere and tends to heat it. Visible light from the Moon also contributes to the effect on Earth's global temperature, but this light is only a millionth as strong as direct sunlight.

Weather on Earth is an extremely chaotic phenomenon. This was publicised in the movie *'Jurassic Park'* with the discussion of the butterfly effect – how a butterfly flapping its wings on one part of the planet can influence the weather far away. Likewise, the tidal effect of the Moon on our atmosphere could lead to tiny shifts in the global airflow and jet streams. These could easily be amplified and lead to larger scale weather phenomenon on Earth's surface. For example, the 18.6 year variation in the Moon's orbital motion has been correlated with a similar periodic fluctuation in arctic ice extent,[111] and an 18.6 year cycle in the temperatures and rainfall on the east and west coast of America has been measured.[112]

### What about earthquakes and volcanoes?

As we learned that the Moon squeezes the Earth and causes the crust and mantle to constantly stretch and deform, does that effect the rate of earthquakes or volcanoes? There is a popular belief that earthquakes are more frequent when the Moon is close to full. But the amount of Moonlight on the

---

[111] "The influence of the lunar nodal cycle on Arctic climate", 2006, H. Yndestad, Journal of Marine Science, 63, 401.
[112] "The 18.6-year lunar nodal cycle and surface temperature variability in the northeast Pacific", 2007, S.M. McKinnel, Journal of Geophysical Research, 112, C2.

surface of the Earth does not affect the Earth's crust in any way.

Most of the volcanic and seismic activity on Earth occurs around the boundaries of tectonic plates, where the immense pressures and temperatures inside the planet are felt at the surface. In the case of volcanoes, the oozing magma below can provide a host of early-warning signals: increased ground deformation, higher concentrations of gases like sulphur dioxide or tiny changes in surface gravity or temperature. But not all volcanic eruptions involve easily detectable pools of magma. Highly explosive 'phreatic' eruptions occur when pockets of superheated, pressurised steam and gas form beneath a volcano, bursting out with little to no warning.

The position of the Moon and Sun has been compared with all of the strongest earthquakes dating back to the 17th century[113]. From over 200 earthquakes above Richter magnitude 8, there was no relationship between the positions of the Moon and Sun and the times of the earthquakes – the times when they occurred seem random. However, the very largest earthquakes, above Richter magnitude 9, tend to occur near the time of maximum tidal stress, but this is based on a just a few earthquakes.

Studies of volcanic activity, either along fault lines below the ocean or terrestrial, do show a small correlation with the lunar tides. Just before a surprise eruption of New Zealand's Ruapehu volcano in 2007, seismic tremors near its crater

---

[113] "Do Large (Magnitude ≥8) Global Earthquakes Occur on Preferred Days of the Calendar Year or Lunar Cycle?", 2018, S. E. Hough, Seismological Research Letters, 89, 577.

became tightly correlated with twice-monthly changes in the strength of tidal forces. In a study of 12 years of continuous seismic data from sensors near Ruapehu's crater, researchers found that in the three months running up to the 2007 eruption, the ground vibrations recorded around the volcano appeared to fall into sync with the effects of the lunar cycle.[114] As the tidal pull of the Moon and Sun increased, so did the seismic activity. This research suggests that signals associated with tidal cycles could potentially provide advance warning of certain types of volcanic eruptions. It will be interesting to see if the same patterns occur for future volcanic activity.

### The effect of the Moon on climate

The temperature on Earth over the past five million years can be determined by measuring the fraction of certain isotopes of water molecules from ancient ice cores and deep sea sediments. Isotopes of water are a proxy for temperature, since heavier isotopes of water evaporate more easily from the oceans when the temperature is warmer.

The deep ice on Greenland and Antarctica preserves a record of the climate at the time the ice formed from falling snow. By taking ice cores from depths of over one kilometre, the climate history over the past million years can be measured. How do we know the ice is that old? Each season's snowfall has slightly different properties than the last. These differences create annual layers in the ice that can be used to count the age of the ice, rather like the growth rings of a tree. The ice contains tiny bubbles of trapped gases, so the carbon

---

[114] "Sensitivity to lunar cycles prior to the 2007 eruption of Ruapehu volcano" 2018, T. Girona et al., Scientific Reports, 8, 1476.

dioxide and methane levels in the atmosphere can be measured. The ice cores reveal a remarkable periodicity during which the temperatures and abundance of greenhouse gases rise and fall with striking rhythmicity.

Compared to the average temperature on pre-industrial Earth, over the past few million years temperatures oscillated by about five degrees centigrade, coincident with carbon dioxide variations from about 190 to 280 parts per million. During this period global temperatures and atmospheric carbon dioxide never reached post-industrial levels.[115]

These long term climate variations resulted in glacial periods that occur with a striking regularity. Over the past two and a half million years, there have been fifty major glacial periods. Initially, the temperature oscillations occurred with a period of about 40,000 years, but for the past million years this period has increased to 100,000 years. But what caused the global temperatures on Earth to rise and fall with such regularity, long before humans were around?

In the 1940's, the Serbian mathematician and astronomer Milutin Milankovitch related Earth's motions to its long-term climate and the ice ages. There are three so-called

---

[115] From one million years ago until the beginning of the 20th century, the average temperatures on Earth varied from about +1C to -8C and atmospheric carbon dioxide by 190 to 280 parts per million. Because temperatures rose followed by carbon dioxide levels, it is thought that carbon was liberated from the oceans. In the past century, temperatures and carbon dioxide levels have increased together. The average temperature of our planet today is higher than at any time during the past million years and is due to a greenhouse effect caused by present day carbon dioxide levels that are above 400 parts per million – also higher than at any time in the past million years.

Milankovitch cycles that lead to periodic variations in the sunlight received on the Earth over timescales of many thousands of years.

The first cycle is due to a slow change in the eccentricity of Earth's orbit over a timescale of 100,000 years. This is due to the varying gravitational pull on the Earth from Jupiter and Saturn. As the eccentricity changes, so does the time it spends near the Sun where it receives more heat.

The second cycle is a slow variation in the tilt (obliquity) of our planet towards the Sun. The Moon is primarily responsible for causing the tilt of the Earth to vary between 22 and 24.5 degrees over a 41,000 year period. It is Earth's tilt towards the Sun that gives rise to the seasons, and these variations can lead to more severe seasons.

The third cycle is the precession of the Earth's tilt, which wobbles around in a circle every 26,000 years – also thanks mainly to the gravitational attraction of our Moon. This precession changes the time at which the seasons occur during Earth's orbit. At the moment in the northern hemisphere, summer occurs when the Earth is furthest from the Sun. But in 13,000 years' time from now, northern hemisphere summers will occur when the Earth is closest to the Sun, which will boost the summer temperatures.

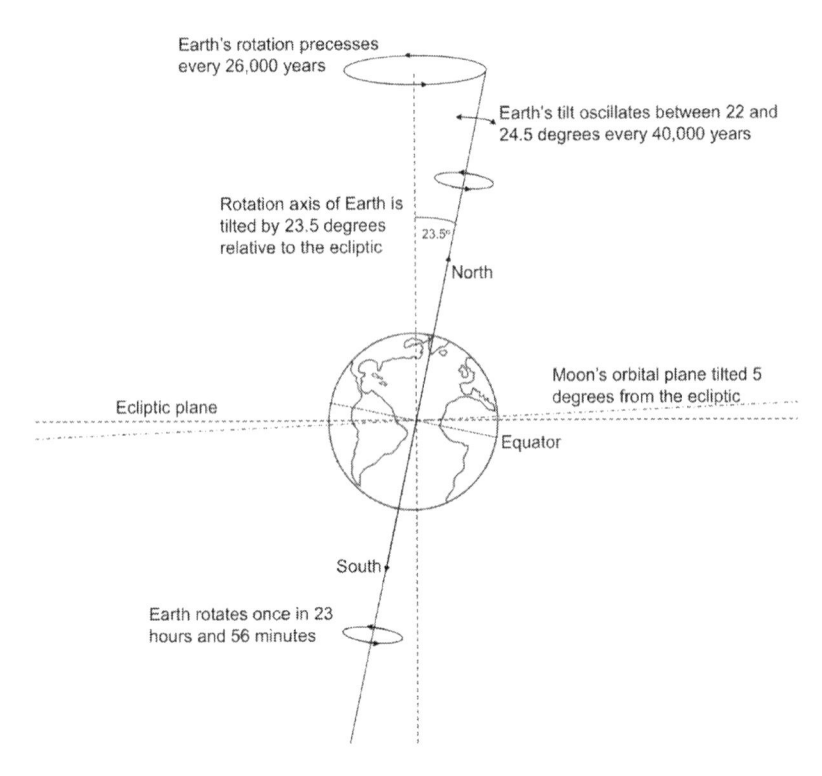

*Our precessing Earth and the orbital plane of the Moon*

The historically measured oscillations in temperature correlate with the Milankovitch cycles, however, these variations in temperature are far larger than expected from the cycles of sunlight reaching Earth.

It turns out that it is rather difficult to calculate how this changing face of the Earth towards the Sun affects the global climate. That's because small changes in solar radiation on Earth can be amplified by many different processes. For example, suppose we are at a time when solar radiation is unusually low in the northern hemisphere. This may cause global temperatures to drop by a degree or less. But this drop in temperature results in larger ice sheets forming, and ice reflects more sunlight than the oceans. This leads to even colder temperatures and even more ice forming. During colder temperatures, more carbon dioxide is absorbed into the oceans, which leads to further cooling and so on, until temperatures have plummeted far below normal values.

The end of the cold period may arise once the Milankovitch cycles lead to a small warming, which melts the ice sheets and releases more carbon dioxide from the oceans. Research from 1997 found that the end of each period of glaciation occurs precisely when the tilt of the Earth reaches its maximum value, which lends support to the ideas of Milankovitch.[116]

The same perturbations that cause the Earth's eccentricity to change also act on our Moon. This leads to variations in the tidal forces on Earth, which could possibly disrupt ice sheets and change the ocean currents and heat flows. The tidal stresses on Earth due to our Moon reach a maximum during the interglacial periods.

Why does the obliquity of our Earth vary? If the Earth were spinning alone in space, it would continue to spin in the same orientation without any variation for trillions of years. But the Earth is slightly extended around the equator as a result of centrifugal forces owing to its rotation. The Moon and Sun, and to a lesser extent Jupiter and Saturn, all pull on the flattened Earth in different directions and on different

---

[116] "Glacial Cycles and Astronomical Forcing", 1997, R.A. Muller et al., Science, 5323, 215.

timescales. The most important of these so called 'gravitational torques' comes from our Moon.

Despite the possibility of our Moon helping cause regular ice ages on Earth, it also has the effect of limiting the variation of obliquity to a small range – saving our planet from more catastrophic variations in temperature on its surface.

I was a postdoctoral researcher at the University of Washington in Seattle in 1993 when I first heard of the new calculations by French astronomer Jacques Laskar.[117] They were presented during a weekly colloquium, and I was transfixed by the findings. Laskar had calculated how our wobbling spinning planet Earth would behave without the presence of our Moon. It was one of the most interesting results that I had heard in my short research career.

Laskar showed that some of the planets in our solar system tip and tilt rather chaotically in response to gravitational perturbations by Jupiter and Saturn. The gravitational tugs from distant planets is tiny, but over thousands of years these effects can add up with dramatic consequences. Today, Mars spins at an angle of 25 degrees relative to its orbital plane around the Sun. However, Laskar calculated that the obliquity of Mars is chaotic, varying between 0 and 60 degrees over a ten million year timescale.

The Earth suffers from the same perturbations, and it should also tip and tilt chaotically like Mars. More recent calculations show that over the past four billion years, the chaotic variations of Earth's spin should have varied between

---

[117] "Stabilization of the earth's obliquity by the moon", 1993, J. Laskar, Nature, 361, 615.

10 and 50 degrees without our massive stabilising Moon, with continual rapid variations of 20 degrees every half a million years.[118] This would lead to climate variations far larger than seen in the geological or ice core records over the past billion years.

However, from its climate records going back half a billion years, we know that the Earth doesn't do this, and Laskar identified that our Moon was the reason why. Our relatively massive Moon pulls on the bulge of the Earth, providing a stabilising force that prevents the obliquity varying by more than a couple of degrees. This helps maintain a relatively stable climate, albeit with the odd ice age, and it has done this for over four billion years. Without our massive stabilising Moon, chaotic climate variations would have regularly occurred, altering the entire evolutionary path of life on Earth.

The presence of our relatively massive Moon had been widely promoted as one reason why intelligent life in our Galaxy seems to be rare – yet another solution to the famous Fermi Paradox. During the seminar about Laskar's work I recall thinking at the time that it would be interesting to calculate just how rare it would be for a planet like Earth to have such a massive Moon. But it wasn't for another ten years that we had the computing power to tackle this problem, and my PhD student calculated that Earth-Moon like systems should be relatively common.[119]

---

[118] "Obliquity variations of a moonless Earth", 2012, J.J. Lissauer et al., Icarus, 217, 77.

[119] "How common are Earth-Moon planetary systems?", 2011, S. Elser et al. Icarus, 214, 357.

# 13. Life in the Moonlight

It has long been thought that the light of the Moon and its mysterious ability to move oceans of water must also affect the lives of creatures that live in its presence. Many of the old stories have turned out to be not true, but there are real effects of our Moon on life that are even more remarkable than the ancient myths.

All living things from bacteria to plants and animals are controlled by biological rhythms. The daily circadian clock cycle linked to night and day is well known, and it is present within nearly all living things. This molecular clock controls activities like gene expression, metabolism and behaviour, such that these biological processes occur at specific times of the day. But there are some organisms with additional biological clock cycles that are shorter or longer than a day. These optimise the organisms' chances of finding mates and avoiding predators and synchronise their maturation and breeding times. Known as circalunar rhythms, these calendar-like clocks are linked to the cycles of our Moon and are embedded deep within the genetic code of numerous organisms on Earth.

### Life in the tidal zone

Life evolves to adapt to its environment, and the tidal zones of the coastline are one of the most complex habitats for life. The sea level rises and falls due to the superimposed lunar and diurnal cycles – the gravitational effects of the Moon and Sun on our spinning planet. Many creatures that live in such zones follow lunar rhythms in reproductive behaviour.[120]

Fiddler crabs always forage for food at low tide when their burrows are uncovered for the longest time period. Their activity is controlled by circatidal 12 hour and 25 minute clocks, which is the time between two low tides. Experiments have shown that even when these creatures are kept in a laboratory under constant light and temperature, the animals are still most active when the tide would be out. Their lunar clock is not activated by light but is built in to their DNA from birth.

There are also examples of marine creatures whose lives are governed by a complex interplay between the daily solar cycle and the monthly lunar cycle. Often the nuptial dances of life inhabiting coastal regions are synchronised to particular days of the month and to specific hours of the night to optimise the chances of finding a mate.

The marine bristle worm *Platynereis dumerilii* always emerges at night to feed, an activity that is controlled by its circadian clock. But it also has a biological clock linked to the synodic lunar month of 29 and a half days. This circalunar clock allows the worm to breed at the same time in the lunar cycle each month – its rhythm follows the brightness of the Moon as it passes through its phases. In recent research, scientists found the specific tiny region in the forebrain of the worms that controls both of these clocks. During the times around the full Moon the worm is less active, perhaps an

---

[120] "Time, Tide and the Living Clocks of Marine Organisms", 1996, J.D. Palmer, American Scientist, 84, 6, 570.

inbuilt mechanism to avoid being seen and eaten by another night-feeding creature. When the bristle worm is kept under constant light levels, its circadian clock stops working but its lunar clock still keeps ticking.

If you think like me that your life is too short then consider the life cycle of the marine midge, *Clunio marinus*. It lives in the intertidal zones of rocky shores along the European Atlantic coast. While the larvae need to be constantly submerged before they hatch, the adults need to deposit their eggs whilst the breeding ground is dry. The evolutionary solution to this conflict has been to drastically reduce the adult lifespan, and to fine-tune adult emergence exactly to the time when the tidal water is as low as possible.[121]

The lowest tides are the spring tides, around new Moon and full Moon when the Sun and Moon are aligned and exert the maximum gravitational squeezing of our planet. The *Clunio* midge uses inbuilt circadian and circalunar clocks that are synchronised to the occurrence of the spring tides. Shortly before the time of the lowest tide, the adult midges emerge and immediately mate, lay eggs and then die in the rising tide. Their entire life cycle lasts just a few hours – evolution in the presence of our Moon can be harsh!

The days of spring tide and the corresponding lunar emergence days of different *Clunio* populations are similar for all places along the coast. However, the exact time of low tide varies at different locations by several hours. That's because

---

[121] "Timing the tides: Genetic control of diurnal and lunar emergence times is correlated in the marine midge Clunio marinus", 2011, T.S. Kaiser et al., BMC Genetics, 12, 49.

of the complex ocean currents next to the irregular coastline. To account for this difference in the lowest tide along the Atlantic coast, the emergence times of *Clunio* populations have been found to be locally adapted to the respective time of low tide. In laboratory conditions, it was revealed that these remarkably precise clocks are controlled by several factors, including the ambient light and temperature levels and the motion of the oceans.

Another stunning example of the effect of the Moon on life takes place one November night each year. Under the light of the full Moon, more than a hundred species of corals simultaneously spawn in Australia's Great Barrier Reef. The corals spew colourful plumes of sperm and eggs like underwater volcanoes. Some species produce just sperm or eggs, but most corals release both together, encapsulated in small spherical packages coloured orange, pink and yellow. At first, the buoyant bundles wait in the lips of corals. Then, in a simultaneous display of colour, numerous corals release their seeds which slowly drift towards the surface of the ocean. Fish, marine worms, and various predatory invertebrates have a feeding frenzy. The ability to correlate this timing is due to the corals having photoreceptors that are sensitive to the light of the Moon as it cycles through its monthly phases.[122]

### Transformations by Moonlight

The legend of the mythical werewolf that transforms during a full Moon dates back to the fifteenth century. Such

---

[122] "Chronobiology by Moonlight", 2013, N. Kronfeld-Schor et al, Proceedings of the Royal Biological Society, 280, 1765.

transformations sound bizarre, but wait until you hear about the Palolo Worm of the South Pacific Ocean. It's about thirty centimetres long and lives in crevices and cavities in coral reefs. As the breeding season approaches, the tail end of its body starts to change. The muscles and organs in its tail degenerate, and the reproductive organs increase in size. There is also a small eyespot in its tail that is sensitive to light. At just the right time, the back end of the animal breaks free and swims to the surface as a separate animal, complete with eyes. The remaining end stays behind and regenerates a new posterior.

This remarkable transformation begins during the last quarter of the Moon in October each year – a few days before the night's sky is dark because of the coming new Moon. One lunar month later it occurs once more during the last quarter of the Moon in November. The freely swimming tails swarm towards the surface of the sea where they release their sperm and eggs. These selected breeding times must lead to the greatest chances of fertilisation and survival of new baby Palolo worms. However, they have not yet adapted to the local Polynesian fishermen who gather the tails in large numbers since they are considered a delicacy to eat.

The daily vertical migration of organisms in the sea is the greatest migration in the world, exceeding in number and mass all the migrations of land animals, birds and insects together. As the light intensity changes, organisms rise and fall in the sea to access food and avoid predators. Typically, creatures move to the surface at night and return to the depths during the day. The connection to light rather than a circadian clock became apparent when their vertical migration was observed to suddenly occur during a solar eclipse. In the Arctic regions during winter the Sun does not rise, and vertical migrations of creatures do not take place, except by zooplankton.[123]

Plankton are all the organisms that drift in the sea, usually microscopic, but they also include larger organisms such as jellyfish. They can be classified as bacteria, fungi, plants or animals. The zooplankton are the animal type, and they tend to feed on the other types of plankton. During the Arctic winter, zooplankton continue to rise and fall when the Moon is visible. Since the Moon rises 50 minutes later each day, the vertical migration takes place over the lunar day of 24 hours and 50 minutes. This is the period over which there are two tides and the time for the Moon to return to the same place in the sky. However, for the zooplankton, they are following the light of the Moon, not the tides. This must optimise their feeding rate. There is a further mass sinking of the zooplankton every 29.5 days in winter, coincident with the periods of the full Moon. This evolutionary adaptation may help them to avoid being eaten by fish and birds that feed during Moonlit polar nights.

### Life in the twilight zone

There are also many examples of creatures far from the oceans whose behaviour is linked to the Moon. Frogs and toads use the lunar cycle and the full Moon to coordinate their gatherings, ensuring that enough females and males come

---

[123] "Moonlight Drives Ocean-Scale Mass Vertical Migration of Zooplankton during the Arctic Winter", 2016, K.S. Last et al., Current Biology, 26, 2, 244.

together at the same time to mate. African dung beetles can navigate far better when the Moon is visible, since they can detect the pattern of polarized Moonlight in the night sky. During this time they can roll their dung balls in a straighter line than on Moonless nights. Perhaps even more remarkable is that they can still navigate during Moonless nights using the light from the Milky Way[124] – the only known creature that uses our galaxy to orient itself!

What about mammals? All bats, badgers and smaller carnivores and rodents are nocturnal. Eighty percent of marsupials and twenty percent of primates are also nocturnal. It is therefore not surprising that many mammalian species are influenced by the levels of Moonlight. And like in marine organisms, it is often a question of finding food or trying not to become food.

On land, some nocturnal animals come out on a well-lit night to hunt, others stay hidden to avoid predators. Other predators prefer Moonless dark nights – it's a tough world if you are another creature's food. A study of Galapagos fur seals found that twice as many were ashore during full Moon than at new Moon. This could be due to either avoiding getting eaten by sharks or the fact that their food is at greater depths during the full Moon. Gerbils avoid eating at night during the full Moon for the good reason that they are more likely to be spotted and eaten by owls.

---

[124] "A Snapshot-Based Mechanism for Celestial Orientation", 2016, B. el Jundi et al., Current Biology, 26, 11, 1456.

In a 2011 study, researchers investigated the eating habits of lions.[125] They found that the belly sizes (food intake) of lions were greatest during the days closest to the new Moon because they had a higher chance of finding creatures to kill in the dark of night. They also analysed the timing of one thousand lion attacks on humans in Tanzania. Two thirds of the attacks were fatal, and the victims were eaten. Sixty percent of victims were attacked in the early evening between 18:00 and 21:45, and attack rates varied strikingly with the phase of the Moon, with most occurring in the weeks following the full Moon, when it rises an hour or two after sunset. In Tanzania, dusk is short and nights are 12 hours long, even during the summer. Hourly attack rates were up to four times higher in the first 10 days after the full Moon, when it does not rise until after sunset. The human kill times were highest in the early evening because humans are still active and outside at these times under a dark, Moonless sky.

Humans have always lived in close proximity to large nocturnal carnivores. Lions were once the most widely distributed mammal in the world – paintings of lions on cave walls in France were made 36,000 years ago. We have long been exposed to the risks of being eaten by lions, a risk that correlates with the waxing and waning of the Moon. The authors of the study on lion feeding habits speculate that this may help explain why the full Moon has been woven into folklore and mythology.

---

[125] "Fear of Darkness, the Full Moon and the Nocturnal Ecology of African Lions", 2011, C. Packer et al., PLOS One, 6, e22285.

**When the sky turns dark**

There have been many anecdotal stories of animal behaviour changing during a solar eclipse, such as an eerie silence from birds which stop singing. But there have also been several systematic scientific studies. In a 1986 study of a group of chimpanzees, the animals were carefully observed for the two days prior to the eclipse and for two days afterwards.[126] Once the eclipse started and the sky began to darken, most of the chimpanzees began to move to the top of their climbing structure. During the period of totality, they all oriented their bodies in the direction of the Sun and Moon with their faces turned upwards. One juvenile stood upright and gestured in the direction of the Sun and Moon. This behaviour did not occur at sunrise or sunset or any other time outside of the eclipse.

Many spiders show diurnal rhythmicity of web-building. For example, Colonial Orb-weaving spiders build their webs at dawn and sit motionless waiting for prey. They then take down their webs at dusk. In a 1994 study of these spiders during a total eclipse, they began dismantling their webs and then at the end of totality several minutes later started to rebuild them.[127]

During the 2017 eclipse that crossed North America, a smartphone app was made available for people to report

---

[126] "Effect of solar eclipse on the behavior of a captive group of chimpanzees", 1986, J. E. Branch and D. A. Gust, American Journal of Primatology, 11, 367.
[127] "Behavior of Colonial Orb-weaving Spiders during a Solar Eclipse", 1994, G. W. Uetz et al., International Journal of Behavioural Biology, 96, 24.

strange occurrences of animal or insect behaviour during the eclipse. There were reports of fireflies emerging, crickets chirping, crabs crawling from their burrows and cows mooing. In a planned study of bee activity during the same eclipse, 16 monitoring stations with microphones were set up in the path of the eclipse next to flowers where bees were active. Before and after the eclipse, bee activity was normal with many buzzing flights past the microphones recorded. At the moment of totality, the bee activity dropped to zero. And as soon as the sunlight emerged again the bees became active again.[128]

These behavioural patterns occur during the sudden transformation from daylight to darkness. Whereas the circadian rhythms of life autonomously govern many bodily functions and behaviours, it seems that the brains of creatures are also influenced by patterns of light and dark. The onset of darkness during an eclipse is being interpreted as the onset of night time.

### Ok, what about plants?

That the phase of the Moon dictates the time to plant crops and when to harvest trees is a myth that originates from the time of Pliny the Elder (who himself may have been influenced by Theophrastus, the Greek successor to Aristotle in the 3rd century BC). Even today there are many myths related to the effects of the Moon on plants, such as the

---

[128] "Pollination on the Dark Side: Acoustic Monitoring Reveals Impacts of a Total Solar Eclipse on Flight Behavior and Activity Schedule of Foraging Bees", 2018, C. Galen et al., Annals of the Entomological Society of America, say035.

common belief that wood from trees is stronger when felled during a waning Moon.

For example, around the first century AD the Greek writer Plutarch stated: *"The moone showeth her power most evidently even in those bodies, which have neither sense nor lively breath; for carpenters reject the timber of trees fallen in the ful-moone, as being soft and tender, subject also to the worme and putrifaction, and that quickly, by reason of excessive moisture; husbandmen, likewise, make haste to gather up their wheat and other grain from the threshing-floore, in the wane of the moone, and toward the end of the month, that being hardened thus with drinesse, the heape in the garner may keepe the better from being fustie, and continue the longer; whereas corne which is inned and laied up at the full of the moone, by reason of the softnesse and over-much moisture, of all other, doth most cracke and burst. It is commonly said also, that if a leaven be laied in the ful-moone, the paste will rise and take leaven better."*[129]

But contrary to the beliefs of many, and despite much research, there is little evidence that plants are affected in any way by the lunar cycle.

There are only a handful of research studies that have reported any observed effect of the Moon on plant behaviour, growth or blossoming. It had already been claimed in the 1920's that the leaves of some plants rise and fall with the night and day of a hereditary circadian clock. From the reanalysis of these old data it was claimed that the amplitude

---

[129] The Philosophie, 1603, p. 697.

of the leaf movement also correlates with the gravitational force from the position of the Moon[130].

The results are not too compelling, since the correlation between leaf rises and lunar phases were done by eye. And the gravitational force of the Moon that is claimed to cause the effect is extremely weak. When the Moon is overhead its gravitational force is about 200 times weaker than that of the Sun and 300,000 times weaker than that of the Earth. With such a tiny force it does not make sense to me that the gravity of the Moon can affect plants on Earth. Researchers in plant biology are also sceptical and have argued that temperature changes and the plants' internal circadian clocks would overpower any measureable effect of gravity from the Moon.

In another study published in 2010, the water content and density of trees felled at different times were correlated with the phase of the Moon.[131] These authors claimed to confirm the statement of Plutarch made almost two thousand years ago. However, in a more detailed follow up study the same year, no significant differences were found between the wood from trees felled throughout the varying lunar phases.[132]

If there were some evolutionary advantage of plant behaviour changing with the phases of the Moon, I would not

---

[130] "Leaf movements and their relationship with the lunisolar gravitational force", 2015, P. W. Barlow, Annals of Botany, 2, 149.

[131] "Looking for differences in wood properties as a function of the felling date: lunar phase-correlated variations in the drying behavior of Norway Spruce and Sweet Chestnut", 2010, E. Zuercher et al., Trees, 24, 31.

[132] "Influence of the Lunar Phase of Tree Felling on Humidity, Weight Densities, and Shrinkage in Hardwoods", 2010, A. Villasante et al. Forestry Products Journal, 60, 5, 415.

be surprised if that manifested itself. But contrary to the numerous animals, insects and marine species whose lives are strongly connected to the Moon, there is little evidence that plants are affected in any way by the Moon.

**But what about humans?**

We evolved from fish that may have had circalunar rhythms, but that was a long time ago and there has been little need for a lunar clock cycle in humans, apart from avoiding getting eaten by lions or other predators. However, folklore runs deep.

In his 1748 text *'Dissertation upon Superstitions in Natural Things'*, the Swiss theologian Samuel Werenfels amusingly describes the superstitious person who reads his fortune in the stars. He, it is said, *"will be more afraid of the constellation fires than the flames of his next neighbour's house. He will not open a vein till he has asked leave of the planets. He will not commit his seed to the earth when the soil, but when the moon, requires it. He will have his hair cut when the moon is either in Leo, that his locks may stare like the lion's shag, or in Aries, that they may curl like a ram's horn. Whatever he would have to grow, he sets about when she [the Moon] is in her increase; but for what he would have made less, he chuses [chooses] her wane. When the moon is in Taurus, he never can be persuaded to take physic [medicine], lest that animal which chews its cud should make him cast it up again. He will avoid the sea whenever Mars is in the midst of heaven, lest that warrior-god should stir up pirates against him. In Taurus he will plant his trees, that this sign, which the astrologers are pleased to call fixed, may fasten them deep in the earth. If at any time he has a mind to be admitted into the presence of a prince, he will wait till the moon is in conjunction with the Sun; for 'tis then the society of an inferior with a superior is salutary and successful."* [133]

By this time in the 18th century, most people in the West believed in the existence of one god, and the nous of the planets and Moon was long forgotten. The association of the Moon with the ability to move the oceans to create the tides led to a belief that the Moon could affect human behaviour because humans were also part water. The 'physics' behind the perceived influence of the Moon had to be due to some unknown force or power that it radiates. But as we have seen in the last chapter, the tides result from the difference between the strength of gravity of the Moon across our planet. That tidal strength across the human body is so small it cannot be measured – even a passing fly exerts a greater gravitational tidal force on your body than the Moon, the Sun and all the stars in the universe combined.

But folklore and superstition can be persuasive, and despite the knowledge of the force that moves the ocean waters, people continued to think that the Moon affects human behaviour. Part of the problem is the coincidence that the female menstrual period is on average the same as a lunar month. Charles Darwin thought that the 28-day human menstrual cycle was evidence that our distant ancestors lived by the sea, and their biological clocks must be synchronised with the tides. However, this seems to be a coincidence. Possums also have 28-day menstrual cycles, but one of our closest relatives, the chimpanzees, have 35-day cycles.

---

[133] Basle, Switzerland. 1748, p. 6.

Despite popular misconceptions that are held even today, human menstrual cycles are not synchronised with the lunar phase, start or end date of the lunar month. And neither do females living together synchronise their periods such that they start on the same date. These beliefs are still widespread, even though they have been confirmed as incorrect by large systematic research studies.

**It's lunacy!**

The full Moon has been linked to many facets of human behaviour, from crime and suicide to accidents and mental health. Some health workers and police believe that they are busiest on nights when the Moon is full, and a few studies have backed this idea, suggesting that crimes, visits to the emergency room and births are highest at those times of the month. To explain this it has been suggested that people are restless and get less sleep during Moonlit nights. But other studies have found no such connections. Researchers matched up police records with the lunar cycle between 1998 and 2003 and found no link between suicides, murders or assaults and the phase of the Moon.[134]

One of the most fascinating lunar myths that still pervades our societies today is that the light of the Moon is linked to madness in humans. The English word for a mad person is a 'lunatic' or someone who has 'lunacy'. The word stems from the Latin word for Moon, 'luna'. In fact, the origins of this myth are not about madness, but about epilepsy. Hypotheses

---

[134] "Influence of Lunar Phases on Suicide: The End of a Myth? A Population-Based Study", 2005, T. Biermann et al., Chronobiology International, 22, 6, 1137.

on the connection between epilepsy, the supernatural and the lunar cycle are as old as civilisation itself.

The oldest known medical handbook is contained within 1000 Sumerian cuneiform tablets dating from around 1500BC. Most are held at the British Museum, and many are still undergoing translation. One of the tablets speaks of a man whose neck turns left, whose hands and feet are tense and whose eyes are wide open. From his mouth flows froth, and his consciousness is fading. This is diagnosed as 'antasubbû', the hand of Sin, who was the Mesopotamian god of the Moon.

In ancient times, many diseases were thought to be due to supernatural causes. Epilepsy was called the 'Sacred Disease', since it was thought to be passed from the gods to those humans who upset the god or goddess of the Moon, Sin or Selene.

Hippocrates of Kos from the 4th and 3rd centuries BC is regarded by many as the father of modern medicine. In his numerous volumes of work he never once appeals to supernatural phenomenon. In fact, he clearly states that this belief must be nonsense, since any supernatural illness would not be amenable to cure via diet or other natural means. The ancient text *'On the sacred disease'* is attributed to Hippocrates, and it is one of the earliest studies on epilepsy. It makes a strong point against a supernatural cause of epilepsy: *"They who first referred this malady to the gods appear to me to have been just such persons as the conjurors, purificators, mountebanks, and charlatans now are, who give themselves out for being excessively religious, and as knowing more than other people."*

Unfortunately, like so many things, ideas became mixed up during the beginning of the Roman Empire. Pliny the Elder was a Roman naval commander, philosopher and author. He wrote an extensive and influential text called 'Naturalis Historia'. It is the largest complete surviving text from the Roman era and is rather like an encyclopaedia of extant knowledge. It treats the folklore and myths related to the celestial bodies as facts. Pliny recounts how the weather, storms, rain and hail are linked to the positions of the planets and the Moon relative to the constellations. An example reads: *"For who is not aware that the heat of the Sun increases at the rising of the Lesser Dog-star, whose effects are felt on earth very widely? At its rise the seas are rough, wine in the cellars ripples in waves, pools of water are stirred...It is indeed beyond doubt that dogs throughout the whole of that period are specially hable to rabies."*

Pliny goes on to state that the Moon corrupted carcases of animals exposed to its malefic rays and induced drowsiness and stupor in those who slept under her beams. He quotes Aristotle: *"Aristotle adds that no animal dies except when the tide is ebbing. This has been widely noticed in the Gallic Ocean, and has been found to hold good at all events in the case of man. This is the source of the true conjecture that the moon is rightly believed to be the star of all life and that it is this star that saturates the earth and fills bodies by its approach and empties them by its departure; and that consequently shells increase in size as the moon waxes, and that its breath is specially felt by bloodless creatures, but also the blood even of human beings increases and diminishes with its light; and that also leaves and herbage are sensitive to it, the same force penetrating into all things."*

Aristotle is often incorrectly credited as the root of the myth of the Moon's effect on human madness. Aristotle, as quoted by Pliny above, did believe that death was connected with the tidal patterns. And this myth extends through the writings of Shakespeare and Dickens and many others. But nowhere did Aristotle write about the Moon and madness.

If the Moon could move the oceans, the prevailing logic was that surely it affects the water inside all of us! And the influential Hippocrates had mentioned that the brain becomes moist during epilepsy. Pliny the Elder wrote: *"The moon on the contrary is said to be a feminine and soft star, and to disengage moisture at night and attract, not remove it. The proof given for this is that the moon by her aspect melts the bodies of wild animals that have been killed and causes them to putrefy, and that when people are fast asleep she recalls the torpor and collects it into the head, and thaws ice, and unstiffens everything with moistening breath."*

Antyllus, a 2nd century AD Greek surgeon who practised in Rome, wrote: *"...the moon rather moistens [the bodies]. And for this reason it makes the brain relatively liquid and the flesh putrid and renders the bodies of people who live in clear cold air moist and dull and, for the same reason, stirs up heaviness in the head and epilepsies."*

The Moon was thought to be intimately related to the cause and relief of diseases. The Venerable Bede, 7th century AD, recounts the tale of a colleague visiting a sick maiden in the nunnery at Wotton, Yorkshire, who lay at the point of death. The bishop inquired when the maiden was bled, and, finding it was in quarta Luna, he said: *"Very unwisely and unlearnedly hast thou done this in quarta Luna, for I remember Archbishop Theodore, of blessed memory, saying that phlebotomy was perilous when the light of the moon and the ocean tide are waxing."*

The paranoia that the light of the Moon induces madness extended until the 19th century before common sense prevailed. In the *'Medical lexicon: a dictionary of medical science'* by Robley Dunglison (1842), the entry for Moon is: *"The moon has been supposed to exert considerable influence over the human body, in health and disease. Such influence has been grossly exaggerated. Not many years ago, it would have been heretical to doubt the exacerbation of mania at the full of the moon; yet it is now satisfactorily shown, that if the light be excluded at this period, the lunatic is not excited more than ordinarily."*

Astronomers of the late 19th and early 20th century were also unconvinced by any effect of the Moon on life on Earth. Here is a quote from George Comstock in his 1903 work *'A Text book on astronomy'*: *"There is a widespread popular belief that in many ways the moon exercises a considerable influence upon terrestrial affairs: that it affects the weather for good or ill, that crops must be planted and harvested, pigs must be killed, and timber cut at the right time of the moon, etc. Our common word lunatic means moonstruck—i. e., one upon whom the moon has shone while sleeping. There is not the slightest scientific basis for any of these beliefs, and astronomers everywhere class them with tales of witchcraft, magic, and popular delusion."*

However, still today, many people believe that there is a connection between human behaviour and the phases of the Moon. There are hundreds of papers published in journals of Psychology and Medicine that claim the Moon influences our behaviour. Many books have recently been written on the topic, such as the 1996 text by Psychiatrist Arnold Lieber: *'Lunar effects: biological tides and human emotions'*.

However plausible an idea you think this is, in all cases when the statistics and correlations are done in a careful way, the conclusions disappear.

*'Does the Lunar Cycle Affect Birth and Deaths?'* was the question posed in a 2006 study of all births and deaths that took place in Australia between 1975 and 2003. During this period there were 7.1 million births and 3.7 million deaths. At a high statistical certainty, no correlation was found between the lunar cycle and fertility, or with the dates of births and deaths.[135]

In 1996 a systematic study was made of all the publications and data that had been used to explore the connection between the lunar cycle and human behaviour.[136] The study found no significant correlation between the lunar cycle and the homicide rate, traffic accidents, crisis calls to police or fire stations, domestic violence, births of babies, suicide rates, major disasters, casino pay-out rates, assassinations, kidnappings, violence in prisons, psychiatric admissions, agitated behaviour by nursing home residents, assaults,

---

[135] "Does the Lunar Cycle Affect Birth and Deaths?", 2011, J. Gans et al., CEPR Discussion Paper no. 532.
[136] "The Moon was Full and Nothing Happened: A Review of Studies on the Moon and Human Behavior and Human Belief" 1996, I. W. Kelly et al., in J. Nickell, B. Karr and T. Genoni, eds., The Outer Edge (Amherst, N.Y.: CSICOP, 1996).

gunshot wounds, stabbings, emergency room admissions, alcoholism, sleep walking or epilepsy.

But still the myth persists. Why?

Observing a bath full of water over one day should be enough to convince anyone that the gravitational tides of the Moon are far too weak to influence small bodies of water.

Studies of human behaviour are often quoted using small numbers of participants. But to gain robust statistical results very large samples are needed. Confirmation bias has certainly played a role in the persistence of the myth. Just as when uncorrelated events confirmed the suspicions of astrologers, when someone observes something strange happening during a full Moon it reinforces the general belief that the Moon affects our lives.

The scientific literature is vast and no one person can keep on top of all the current developments. Anecdotes passed on from friends or in popular magazines and newspapers are likely have more effect than a research study that claims a null effect. I think that a large part of the persistence of ancient myths is also due to media bias. How often do you read a report in the newspapers such as 'a glass of wine a day is good for you' or 'chocolate makes you lose weight'? Most often this is because of journalists misrepresenting the results of researchers, or they are trying to sell a story playing on what people want to hear.

Here is a typical example of media bias I found. A 2013 study led by Christian Cajochen from the Psychiatric Hospital of the University of Basel analysed the sleeping patterns of people over a week.[137] The researchers found that during the full Moon the subjects took five minutes longer to fall asleep and slept for an average of 20 minutes less. This supported a widely believed myth that humans are restless during the full Moon. The study was reported in the news and across the internet as evidence for a lunar effect on humans. And not just the 'dailys' but respectable news sources and magazines such as the *'BBC'*, *'Scientific American'*, *'Science magazine'*, *'New Scientist'*, *'CNN'*, *'National Geographic'*, the *'New York Times'* and many more.

What was not widely reported was that the study only monitored 33 participants. And what was also not reported on were the results of several far larger and recent studies. In a 2017 study led by researchers from the Helmholtz Zentrum München, the activity and sleep patterns of 1,411 Germans were monitored between 2011 and 2014, and no correlation with the Moon phase was observed.[138] In another medical publication, 2,328 individuals from five different countries were monitored for sleep patterns over time, and again no correlation with the Moon phase was found.[139] Neither of these results were reported by the media. Clearly, myth

---

[137] "Evidence that the Lunar Cycle Influences Human Sleep", 2013, C. Cajochen et al., Current Biology, 23, 15, 1485.

[138] "Physical activity, subjective sleep quality and time in bed do not vary by moon phase in German adolescents", 2017, M.P. Smith et al., Journal of Sleep Research, 26, 371.

[139] "Evidence for daily and weekly rhythmicity but not lunar or seasonal rhythmicity of physical activity in a large cohort of individuals from five different countries", 2015, R. Refinetti et al., Annals of Medicine, 47, 530.

busting is not what readers want to hear, or at least that is what the journalists believe.

## 14. The Future

We began this journey of the discovery of our Moon by recounting some of the dreams of humanity to reach for the Moon. Just as those who enabled the lunar landings were inspired by stories and imaginations, the Apollo program inspired a new generation of scientists and explorers with dreams of returning to the Moon or even venturing further into space. So what does the future hold in store for our Moon?

The Moon landings happened because of one reason alone, to beat the Soviet Union. Once that task had been accomplished there was little discussion of humans returning to the Moon. I find it rather sad that despite all the technological advances since the 1960's, humans have not been further into space than the few days it takes to reach the Moon. Whilst there has been big talk of Mars as the next destination for humans, I think that is unlikely to happen in my lifetime at least.

In the last decade several unmanned missions to the Moon have taken place, and interest is again rising in returning to the Moon, both as a preparation for more bold ventures and for commercial, private and scientific purposes. In the next decade there are over ten missions scheduled to visit our Moon, from the American, Japanese, Chinese and Russian national space programs. These missions are all unmanned orbiters or remote sample and return missions. Only the private companies, such as SpaceX and Blue Origin, have announced that they will send passengers around the Moon within the next decade, although no firm launch dates have been given.

In 2013 the Chinese space agency successfully carried out the first controlled landing on the Moon since the Soviet Luna 24 mission in 1976. At the beginning of 2019, the Chang'e-4 mission successfully placed a robotic lander on the far side of the Moon to explore the giant Aitken impact crater and measure the chemical composition of its rocks. It also successfully germinated the first plants on the Moon, cultivated within a mini greenhouse. A second mission, Chang'e-5, is planned for launch in 2019 and will land on the northern region of the Moon, where the lunar rocks are thought to be the youngest. This robotic lander will dig two metres below the surface and bring two kilograms of Moon rocks back to Earth.

The Chinese National Space Administration has fairly realistic objectives with well-defined goals. In the coming years they wish to establish a crewed space station, followed by landing humans on the Moon and then the creation of a permanent Moon base, followed in the longer term by unmanned and then manned missions to Mars. This all sounds great, but beneath this lies the worry that the Chinese are developing their space program primarily for military applications.

In the last few years, nearly every space faring country has independently announced plans to land astronauts on the Moon and establish permanent Moon bases. In 2015, the Russian space agency Roscosmos stated that Russia plans to place a cosmonaut on the Moon by 2030. Roscosmos has stated its aim of establishing a permanent Moon base and is leaving Mars to NASA to avoid any future costly space race.

The Russians seem to have stimulated a new space race, since the European Space Agency has also proposed 2030 as the time when they will construct a permanent Moon village, occupied at first by around ten astronauts, but expanding to 100 people by 2040. There even seems to be interest in a collaboration with China to achieve this goal. In 2017 the Japanese space agency, JAXA, revealed its plans to put a human on the Moon by 2030. Even private space companies have stepped onto the lunar village bandwagon.

NASA has also recently announced its intention of returning to the Moon, and in 2017 began a cooperation with Russia to establish an orbiting lunar space station as a first step. Called the Deep Space Gateway, it is intended as a base for further exploration of the lunar surface and the space beyond our Moon.

But despite the grand claims of many national, international and private space programs, these intentions of sending humans to the surface of the Moon have not proceeded much further than the proposal stage. Even the cost of returning to the Moon is not known. A short lunar visit should cost less than the 120 billion dollar Apollo program, but have experience and advances in technology reduced that cost significantly? A 2009 study by NASA estimated a return mission to the Moon would cost at least 50 billion dollars of funding on top of its current budget. Although a more optimistic 2018 analysis claims that humans could return to the Moon for a tenth of the cost of the Apollo program –

around 10 billion dollars – the reduced costs are anticipated to come primarily from collaborations with private companies.

A Moon base will cost far more, although no one has actually carried out a detailed cost estimate. Some have claimed 50 billion dollars, but I can only guess that it would cost a comparable amount to the international space station's 150 billion dollars. So far, no government has pledged this scale of funding, and no private individual has that much money. Perhaps this could be funded via space tourism, selling Moon collectables and a pay per view reality show with contestants sent to the Moon. Ok, I'm half joking about the last idea.

The most experienced and largest funded space program is NASA, and unfortunately the goals of NASA are at the whim of whomever is the current president. In 1984 Ronald Reagan wanted to develop the space military program. George H.W. Bush had a vision of returning to the Moon. Between 1993 and 2001 Bill Clinton cut NASA's budget every year and made the focus the International Space Station. In 2004 George W. Bush reinstated his father's vision and gave NASA the mandate of returning to the Moon by 2015. President Obama changed the focus to landing humans on an asteroid and then Mars. Now we have President Trump, who has changed the focus of NASA back to the Moon despite cutting funding for already planned lunar missions.

If we wish to return to the Moon and secure the funds to do so, there needs to be good motivation for such a project. And there is. In 2006 NASA compiled a list of 180 different reasons for establishing a human presence on the Moon. Amongst the topics discussed in this list were astronomy, Earth observation, understanding lunar geology and its origins, material science, human health studies, environmental hazard mitigation, development of life support systems and space habitats, assisting further space navigation, mining lunar resources, creating lunar heritage sites, developing lunar commerce and tourism, public engagement and inspiration.

All of this would be best accomplished by establishing a permanent Moon base, occupied by scientists, astronauts and maybe the occasional space tourist. So what would a Moon base look like and what could we really achieve by returning to the Moon?

**A village on the Moon**

The idea of a lunar colony stretches back to the earliest science fiction of the 17th century. In 1640 Bishop John Wilkins wrote *'A Discourse Concerning a New World and Another Planet'*, in which he predicted a human colony on the Moon. Tsiolkovsky and others made serious suggestions of a Moon base at the beginning of the 20th century, and prior to the Apollo program there were many creative ideas from both science fiction and scientists.

In 1954 Arthur C. Clarke proposed his Moon base concept and envisaged various inflatable lunar modules that could be covered over with moon dust. The modules would resemble igloos and come equipped with radio masts, algae based air filters and even nuclear reactors. He suggested using enormous electromagnetic cannons to 'fire' supplies to orbiting space ships. His 1968 work on the movie '*2001: A Space Odyssey*' and concurrent book takes his ideas further. The Moon base is mainly constructed beneath the surface to protect it against solar radiation and micro-meteoroid impacts.

At the beginning of the Moon race, both the United States and the Soviet Union had plans for a crewed Moon base. These were intended for both scientific and military purposes. The United States' 'Project Horizon' aimed to place twelve soldiers in a Moon base by 1966. Over 100 launches of the Saturn A-1 rocket were proposed to shuttle over 300 tons of components to the Moon. The cost was six billion dollars at the time, which in today's money is over 50 billion dollars. The plan was ultimately rejected by President Eisenhower.

The Soviet Moon base was intended to follow their manned Moon landing program. Components were to be delivered robotically in advance of the dozen or so cosmonauts who would live there. Plans began in 1960 and were later approved by the Soviet government. Over fifty tons of equipment were to be ferried to the Moon, including a 21 ton habitation module with 60 square metres of laboratory and living space. The project was cancelled in 1974 after the end of the successful Apollo program and the realisation that America had dropped its goals of a Moon base.

In 2016 the Moontopia competition was held to ask architects and designers to visualise life on the Moon. The winning design involved a gradual colonisation of the Moon, using a combination of 3D printing and self-assembly. Each structure would be made of carbon fibre and based on origami. Astronauts would initially assemble the base, with future provisions being made for space tourism and colonies later on. It sounds more like a complicated Ikea construction kit – fun to assemble in a space suit! The concepts were all very creative, but many of the designs were in the realm of science fiction, such as an orbiting space station with a space elevator leading to the surface.

Sending supplies from Earth to construct a Moon base would be expensive, but a lunar settlement could be constructed using locally sourced material and energy. The lunar highlands are abundant in anorthosite, which is a rock made of calcium, aluminium, silicon and oxygen. By smelting the mineral you can release many of the rare ingredients necessary to construct your Moon base. The mineral troilite, an iron-sulphur compound rare on Earth, is also present in the lunar crust. Experiments recently carried out on Earth show that the sulphur from troilite can be extracted and combined with lunar dust to produce a building material stronger than cement. Although melting rocks into their components takes a lot of energy the Moon has plenty of unobscured sunlight for solar power.

One of my favourite science fiction movies is *'Moon'*, directed by Duncan Jones. A lone astronaut spends a three year shift on the Moon managing, the mining facilities that extract Helium 3 from the lunar surface to be sent back to Earth for use in fusion power generators. The Chinese Lunar Exploration Program is investigating the prospect of lunar mining, specifically looking for the isotope helium-3 for use as an energy source on Earth. Often mentioned in science fiction, I think helium-3 mining will stay in the realm of science fiction for quite some time. We don't know how much, if any, exists on the Moon, no one has estimated the cost of mining it and returning it to Earth, and we do not have a use for it at the moment since fusion power is some decades away.

Any Moon base must provide its own atmosphere and protection against micrometeorite impacts and the high speed spray of ejected debris from larger impacts. With the detailed knowledge we now have of the surface of the Moon, the designs of a possible Moon base have changed. We know that the surface is solid and can be walked upon. There is evidence of numerous ancient lava tubes on the Moon. A Moon base inside the Moon, using these natural passageways, would protect lunar inhabitants from the extreme changes in temperatures and would shield against small meteorite impacts.

However, the presence of frozen water ice at the poles of the Moon makes these regions even more likely places to establish a permanent lunar home. Water is going to be the most valuable commodity for any Moon base. On the North Pole of the Moon there is an estimated 600 million tons of water-ice. On the poles of the Moon there are also craters that are in permanent darkness, never receiving the light of the Sun. And there are lunar mountains nearby that are illuminated by sunlight over ninety percent of the year.

A Moon base deep within a crater would be partly shielded from asteroid strikes and would have a practically unlimited source of solar energy nearby, as well as access to plenty of frozen water. Sunlight could be reflected via giant mirrors into the crater depths, allowing an 'Earth-like' night and day environment to be artificially created.

Water ice can be collected from the craters at the lunar poles or extracted from the lunar soil. Solar panels on the poles of the Moon receive the full kilowatt per square metre continuously for as much as 90 percent of the year. This energy could be used to split the water molecules by electrolysis or artificial photosynthesis, generating rocket fuel and oxygen to breathe. Establishing a lunar base from which to launch deep space missions would massively increase the payload to fuel ratio, thanks to the Moon's lower gravity. This would allow future exploration of the solar system at a fraction of the current cost.

**Science on the Moon**

A permanent Moon base would be incredibly useful for science. Many manufacturing processes and experiments on Earth are carried out in vacuum-like conditions. Expensive equipment is needed to reduce the air density, but even the best vacuum we can produce on Earth still contains about a million molecules per cubic centimetre. To create a vacuum for an experiment on the Moon one only has to take a container and open it outside.

The Moon would provide perfect conditions for astronomical observations, especially if you had a telescope in the permanent darkness of one those craters that never received sunlight. With no light pollution on the Moon, nights are truly dark. With no atmosphere there is no diffraction of light that causes the twinkling of stars and limits the resolution of Earth-based telescopes. Although orbiting telescopes offer the same advantage, they are small in size compared to telescopes on the ground and they are difficult and expensive to service.

Large optical telescopes could survey the atmospheres of exoplanets to search for the biosignatures of life upon their surfaces. The optical precision without a blurring atmosphere would allow us not only to easily detect worlds orbiting other stars, but to find out if they have Moons of their own. Infrared astronomy would be particularly interesting, since on Earth these wavelengths are blocked by our atmosphere. And the cold night time temperatures would be ideal for limiting noise in our electronic detectors.

Radio astronomy on Earth is plagued by the background radio noise, not only from broadcast radio, but from all the devices that use radio waves to communicate, from garage door openers to microwave ovens. A radio telescope on the far side of the Moon would be shielded from radio noise, allowing a clean view of the universe. It would be ideal for searching for extra-terrestrial intelligent life in our galaxy via their radio transmissions.

The Moon would also be an ideal location for dedicated telescopes that search for dangerous Near Earth Objects. Observatories on the near side of the Moon would have a continuous view of the entire disk of the Earth, allowing us to monitor global effects of climate change, the reaction of Earth's atmosphere to solar activity, changes in snow cover and glaciation and the condition of our oceans and marine ecosystems.

### Exploration of the Moon

After centuries of speculation and scientific thinking we have some reasonable ideas of how our Moon formed, but there is no agreed upon model for its origin, and there are many remaining mysteries about our Moon. To unlock further secrets held by our Moon we would need to visit its surface once more. The samples collected five decades ago by NASA's six successful Apollo missions are very valuable, but the collection is just from six landing sites, all on the near side of the Moon and all close to the equator. The Moon has a diverse geology across its surface, and the Apollo collection is not a large or complete sampling of the whole Moon. At a recent scientific conference on the Moon, over fifty interesting locations were identified by scientists from which samples should be collected and analysed.

All of those early lunar missions, manned or unmanned, only returned rock samples found on, or just beneath, the surface. The samples were collected from the dusty layer of pulverized lunar rock — the so-called regolith, a product of meteoritic bombardment. We don't have any samples of the bulk interior of the Moon, just those that have been scraped off the present lunar crust. This is not sufficient to understand the composition of our satellite and solve the riddle of its origin. As we have already learned, these samples all suggest that the Moon is made of the same material as the Earth, but we are not comparing the bulk compositions. It is the composition of the Moon's mantle, its interior, that really counts, not just the thin dusty layer on the surface. Our effort so far is rather like trying to understand the composition and history of the Earth by scooping up some sand from the Sahara desert.

The confirmed discovery of water on the Moon has raised some basic questions, such as how it got there and how it survived the high temperatures that resulted during the favoured impact origin for our Moon. Perhaps it originated from comets and asteroids which vapourised upon impact, condensing at the bottom of cold craters on the Moon's poles. Another suggestion is that hydrogen from the solar wind combined with mineralised oxygen in the lunar regolith, forming water which escaped and found its way into space or got trapped in the cold crater bottoms. Or perhaps it came from the Earth, surviving the initial giant impact and condensing onto the Moon shortly after its formation.

Then there is the puzzle that of more than 2000 Moon rocks brought back to Earth, many have been radiometrically dated to around 3.9 billion years ago, when they presumably formed from the melted surface material during several large impact events. Very few of the Moon rocks were older than that. What happened at this time on the Moon? Some scientists have suggested that the ejecta from the giant Imbrium crater are scattered all across the Moon, and the Apollo missions to nearby regions simply collected the debris from that one giant impact. The impact rate over time on the Moon, and on all the solar system planets, relies on knowing the accurate dates of many lunar craters. If we have only measured the date of one crater then our knowledge is far from incomplete. The only way to figure out the early history of the Moon is to analyse samples from many different regions and craters.

After the last of the giant impacts on the Moon had taken place, volcanic eruptions filled some of these basins with lava and created the formations of the current lunar landscape. The luna maria mostly exist on the near side of the Moon. There are also many impact basins that were never flooded with lunar lava, almost all of them on the far side of the moon. Why is that? Although there are several suggested scenarios for this strange difference between the near and far side of the Moon, only a thorough exploration of the Moon will reveal its secrets.

Most of the geological activity on the Moon occurred early in its history. However, scattered across the near side of the Moon are numerous small patches, up to five kilometres across, that look like recent deposits of basalt rock. Some of these patches must be relatively new, since they are not covered by small impact craters. They have steep boundaries which would also not withstand the rain of debris onto the Moon if they were ancient features. This implies that lunar volcanism may have been ongoing until as recently as 100 million years ago, which would turn our ideas on the internal structure of the Moon upside down. From the transmission of seismic waves through the Moon's interior we think most of the mantle is cold solid rock. In that case, where did the recent basalt deposits come from? There is nothing like them on Earth.

There are also continual surface changes on the Moon as a result of the crust deforming and cracking. As the Moon continues to cool its inner regions contract. The brittle crust responds to this by fracturing, resulting in thousands of narrow 'scarps' that are too thin to have survived for long on the surface. These fault like features are thought to have occurred in the last 50 million years. The orientation of these features suggests that Earth's tidal forces may contribute to their formation.

**Would a Moon base be fun?**

I'm sure there would be plenty of people who would love to spend a few days or weeks on the Moon. What an experience! I would certainly prefer to spend time on the Moon, with its magnificent view of our Earth, rather than far-away Mars. But would you want to spend a year on the Moon away from most of your Earthly comforts? In your sealed Moon base buildings, you could move around as in any hotel room, but to venture outside you would need an expensive and sophisticated space suit with all of its life support system functions, from supplying oxygen and removing vapour and condensation to maintaining a bearable temperature.

For the passionate mountaineer, just think of all those first ascents of kilometre high mountains you could achieve. With gravity being one sixth of the Earth's you could leap and bound to the tops of the highest peaks. I'm sure the future Lunar Olympics will be fascinating to watch. Imagine all the world records you could break!

Satellites could certainly provide high speed internet to the Moon. And the light travel time from Earth to the Moon is just 1.3 seconds, so real time conversations are possible although with an annoying lag. Rather like overseas telephone conversations when we were kids. Computer games against opponents on Earth, which need fraction of a second communication times, will certainly not be fun.

You might miss your favourite restaurant on the Moon, since local produce will be hard to come by. On the Moon there is no atmosphere to provide carbon dioxide for plants, no bacteria to help feed their root systems, and the temperature changes are extreme. However, large green houses could be used, with solar shields and reflectors providing a controlled environment with essential minerals recycled from human waste. The lunar soil does contain most of the nutrients necessary for plants. In 2014 researchers demonstrated that within a closed temperature controlled environment, plants can indeed germinate and grow within a rocky powder that closely matched the lunar regolith.[140]

The Soviet Union began a series of experiments in a Siberian facility, starting in the early 1960's, to study the feasibility of sustaining human life inside a small closed ecological system. The first BIOS-1 experiment enclosed a human in a 12 cubic metre space connected to an algae tank that supplied oxygen and removed carbon dioxide from the atmosphere. A fully closed system took over a decade to design, and the sealed BIOS-3 experiment eventually supported three crew members for six months.

In one experiment where no food was imported from the outside, it took about one hundred square metres of growing space to feed three people and supply their oxygen needs. Artificial lights were used, and the crew carried out all tasks with no help from the outside. A human needs about one kilogram of oxygen to breathe each day, and 30 square metres

---

[140] "Can Plants Grow on Mars and the Moon: A Growth Experiment on Mars and Moon Soil Simulants", 2014, G.W. Wieger et al., PLOS One, e103138.

of plant growth is more than sufficient to supply this. In fact, the oxygen levels would have been dangerously high, so the inedible biomass was burned to produce extra carbon dioxide for plant growth and to maintain a safe oxygen level inside the experiment. All water was recycled and the 'waste products' from the crew were used as plant nutrients. The only products that were brought in from the outside at the start of the experiment were sanitary products, salt and some additional plant nutrients – phosphorous, nitrogen, sulphur, potassium and magnesium.

In the same experiment, wheat, barley, peas, turnips, carrots, cucumbers and tomatoes were amongst a dozen species studied under lunar conditions of night and day. During the 15 day dark period, temperatures were reduced to three degrees centigrade. This maintained plant vitality, and most species survived the dark period and grew during the 15 days of light. Some crops did not produce as much crop volume, such as carrots and turnips with yields 50 percent down. Others, such as beet, fared even better than outside under these conditions. Only the tomatoes and cucumbers failed to flourish and died, so no pizza or salads on the Moon. Surprisingly little research on plant adaption to the Moon has been carried out since.

It is not known how crops like these would respond to the lower gravity of the Moon. On Earth plants 'know' to grow upwards because of the presence of tiny mineral grains that sink to the bottom of certain cells. These stimulate the release of hormones that cause plants to grow upwards. It will be interesting to see if the gravity on the Moon is sufficient to give the plants a sense of direction, rather than just growing in random directions across the surface.

Some food and supplies will certainly have to be flown in from Earth. At today's prices, it costs about 10,000 dollars to launch a kilogram into space. So a bottle of wine is not going to be cheap on the Moon. For travellers on a budget, self-sustaining vats of bacteria could be used to provide the basic nutritional food you need, but it might get a little boring after a while consuming those little green cyanobacteria pills. And I hope the Moon base has a fully functioning hospital lab, because if you have a nasty toothache on the Moon it might be several weeks before the next flight back to Earth is planned.

So a Moon village may not be enough – why not plan a Moon city, enclosed in a giant protective dome, with streets and houses, shops and bars, theatres and restaurants, daily flights to and from Earth and regular excursions to see all the sights of the Moon? Those buildings with a permanent view of the Earth would be prime real estate – what a sight to see the changing weather on Earth and the continents slowly moving past. There are over 2000 billionaires on Earth today with a combined wealth of 10 trillion dollars. They could build their own gated paradise on the Moon for a fraction of their total savings. It would be a safe haven from which they could watch the demise of our planet as their money making industrial activities turn its surface into a second Venus.

But seriously, perhaps space tourism alone could fund such a venture. If the cost of a weekend vacation on the Moon was a few hundred million dollars I'm sure there would be enough takers to fund a large part of the initial development costs. Space tourism is already a thing. Tickets to the International Space Station have been sold to eight private individuals for around 20 million dollars each. SpaceX has already sold tickets for a trip around the Moon and back.

In 2018 the Japanese entrepreneur and art collector, Yusaku Maezawa, prepaid SpaceX for a personalised trip around the Moon. That trip is planned to take place in 2023. Maezawa wishes to use the experience to inspire several artists, who will accompany him for free to portray their experiences of the journey to the Moon. The exact price of the trip is not known, but Elon Musk said it was roughly the same as sending a crew to the ISS. NASA currently pays Roscosmos about 70 million dollars per seat.

A space tourism firm already works together with Roscosmos to send private citizens into space using Soyuz space craft. They are advertising a two seat journey around the Moon to be launched in the early 2020's. The cost of this journey has been quoted as 175 million dollars per seat. There are currently at least half a dozen commercial companies who plan to offer space tourist flights in the very near future. Richard Branson has already sold around 700 tickets at 250,000 dollars each just for a suborbital flight to the edge of space above Earth.

If you could afford such a trip, would you take it? I would certainly love to fly around the Moon, or walk upon its surface. But travelling in a rocket is similar to sitting on the end of a large firework. It's uncomfortable, dangerous, the food sucks, the internet isn't great, and there are many things that could go wrong. If we can progress as a civilisation, rocket travel will probably become an everyday thing. We are just alive at the time when it's all a little bit experimental. And that says a lot about those bold pioneers who made the first trips to the Moon.

### Preparing for destination Mars

Several private space companies have made even bolder claims than returning to the Moon and have announced that they will place a human on Mars by 2030. Whilst landing humans on Mars is a grand goal that humanity should one day achieve, it would be an incredibly expensive mission and will only work with international collaboration. I do not expect to see this happen in my lifetime at least. But I do hope and believe that we will see a permanent base on the Moon being established within the next two decades, and that would be an ideal way to prepare for bolder ventures beyond.

A lunar base would provide invaluable experience in developing and operating life support systems, such as sustainable energy sources, food supplies and recycling water. Although some argue that going back to the Moon before Mars will divert funding and delay a Mars habitation by decades, I think it is essential for developing the necessary technology. The trip to Mars will be at least ten times the cost of establishing a Moon base. And if we can't establish a base on the Moon, we don't have a chance of doing that on Mars.

The steps for establishing a lunar base are similar to what is needed for Mars. But the Moon is just three days not half a year away, so troubleshooting and help is within easy reach. It would allow us to study the effects of space and reduced gravity on the human body and to design life support systems for the long journey to Mars. It would also be far easier to reach Mars from the Moon, rather than setting off from the Earth.

To overcome the pull of gravity and reach another body in space you need to achieve a certain speed. A journey to Mars from Earth's surface requires a minimum rocket speed of 13 kilometres per second. The first stage of the Saturn V rocket required a million litres of fuel just to reach beyond Earth's atmosphere. Due to the Moon's weaker gravitational field, the same journey from the lunar surface would only require a rocket speed of about three kilometres per second, roughly one third of that necessary to reach the International Space Station from Earth. Even if the Mars rocket is constructed on Earth, it could gather most of its fuel via a brief stop in a lunar orbit, where it could be supplied from the lunar surface.

A successful interplanetary and eventually interstellar space program is a grand and achievable goal. The next big frontier of space is indeed walking on Mars. And if humans are to survive in the future I agree with those who argue that eventually colonising other worlds should be a long-term, thousand year goal of humanity. The counter-argument that is frequently made is that we have big problems on Earth to solve first. That is true. But there is no reason that both cannot be achieved, and having a space program does not mean that we stop working towards a safer, greener Earth. The technological developments that arise from pushing new frontiers cannot be predicted but always lead to new breakthroughs. That's one reason why governments fund such a breadth of science – no one knows where the next major life-changing discovery will come from.

We are certainly capable of reaching beyond the Moon, but is it a question of money or priorities?

The cost of the Apollo missions in today's money totalled 120 billion dollars. Sending humans to Mars is estimated to cost several times this, despite the improvements in technology. No one country is investing this much money in this objective. Elon Musk has a net worth of around 20 billion dollars, so he cannot afford this alone. Neither does he have the resources – the Apollo program took several hundred thousand people and collaborations with tens of thousands of companies and universities. SpaceX has just 7,000 employees.

Destination Mars would take the combined resources and funding of all the world's active space programs. And not just for one year, it would require a long term collaboration that would span many years.

If you think that all this is a lot of money for space, then you are correct. But in comparison, the development of the F-35 fighter jet cost 332 billion dollars, and it will cost another one trillion dollars to maintain. Each year the United States spend more on military than the entire Apollo, Space Shuttle and International Space Station programs combined. And each year the rest of the world combined spends a similar amount on the same wasteful resources for war.

Globally, we spend about two trillion dollars per year on the military, and wars have an indirect cost that is estimated to be as high as a staggering 14 percent of the GDP of the world. In 2015, a careful cost analysis of the effect of war came to 13.6 trillion dollars. On the other hand, peace building activities were funded at a level of 8 billion dollars, less than one percent of cost of war. In the same year, space programs totalled around 50 billion dollars, most of that going into the International Space Station and Earth monitoring programs.

**Who owns the Moon?**

Before we build our lunar village or city, we should check our property rights. Does anyone have the right to build a Moon village anywhere they like on the Moon? Can you claim property rights for your Moon villa with the best views of Earth? Could you go and take a joy ride in one of the lunar rovers left behind by the Apollo mission?

It would be fascinating to study the lunar landers that have been standing on the Moon for fifty years. The impact of being exposed to space for so long would allow us to quantify the effects of the bombardment by micrometeorites and cosmic rays on space craft materials. However, in 2011, NASA requested a 500 metre 'keep-out' zone around their historic lunar landing sites.

One of the best known pictures from the Moon shows Buzz Aldrin standing next to the freshly planted American flag. Less than a century before, planting a national flag on a patch of Earth was testament to claiming that land for your sovereign country. Does that mean that America owns the Moon? The United States knew at the time that this act would raise major political issues, but planting the American flag was intended as a symbol of their achievement, not an act of claiming the Moon.

Already in 1967 the United Nations sponsored the 'Outer Space Treaty', establishing all of outer space as an international commons and the province of all mankind, explicitly forbidding any nation from claiming territorial sovereignty of the Moon or any other cosmic body. It also prohibits the use or testing of weapons of any kind on the Moon or in space. This treaty has been signed by 102 countries, which included America and all those nations with an active space program.

However, since then private companies have emerged and may not be covered under some existing treaties since the legal interpretation can be complicated. This has led many individuals to claim the Moon and even sell plots of land, by arguing that the Outer Space Treaty only applies to nations and not individuals. However, if you look at the actual document you can see that this is not the case: *"States Parties to the Treaty shall bear international responsibility for national activities in outer space, including the moon and other celestial bodies, whether such activities are carried on by governmental agencies or by non-governmental entities, and for assuring that national activities are carried out in conformity with the provisions set forth in the present Treaty. The activities of non-governmental entities in outer space, including the moon and other celestial bodies, shall require authorization and continuing supervision by the appropriate State Party to the Treaty."*

Whilst a more detailed treaty written in 1979 called 'The Moon Agreement' gives more detail on the property rights and usage of the Moon, it has only been signed by 17 nations, all of which are minor players in space exploration.

A United Nations treaty is technically binding, but there is no department of 'Space Police', which means it is difficult to enforce. There is usually international pressure by other nations when a country strays from the principles. There have been some debates over the years about some of the major principles of space law. For example, the Outer Space Treaty does not explicitly address the commercial exploitation of natural resources on the Moon or other celestial bodies such as asteroids.

The reality of it is, some countries just do whatever they want regardless of existing treaties. The Strategic Defence Initiative, also known as the Star Wars program, was announced by United States president Ronald Reagan in 1983. Weaponised space satellites were studied and planned, but fortunately this was concluded to be too technologically demanding and expensive. In 2015 president Obama signed into law the United States Space Act, which permits private firms to carry out activities that may go against international law, such as mining asteroids for resources. It goes against the Outer Space Treaty, which prohibits countries from appropriating any part of outer space.

**Terraforming the Moon**

Whereas a Moon village may happen within the coming decades, humans would have to live in an environment sealed from the outside vacuum of space and protected from micro-meteors and high energy cosmic rays. Venturing outside would always require clumsy space suits and carrying your own oxygen to breathe. Could we change that by terraforming the Moon, creating a protective and breathable atmosphere and allowing plants and animals to live on its surface?

The Outer Space Treaty of 1967 and the Moon Agreement of 1979 have little to say on the issue of terraforming our Moon. It is not impossible, but it would be immensely difficult. The technology needed is centuries away, yet in a few thousand years from now perhaps it would be desirable to have a second home upon the Moon in an environment that you could walk around without a space suit.

For over four billion years our Moon has been no place for life. Although we do not yet have a complete picture of how life began on Earth, we do know many of the conditions that are necessary for life to emerge and thrive. And the Moon lacks most of these basic features. Because of its small and mainly solid metal core, the Moon lacks a magnetic field, which acts as a protective shield against high energy particles from the Sun. Any gasses that emerge from the interior are quickly swept into space. Although life does not need an atmosphere to emerge or thrive, an atmosphere is essential for maintaining a habitable climate in which liquid water could exist on the surface of the Moon. A lunar atmosphere would prevent high energy cosmic rays from penetrating to the surface. These particles can damage cells and cause radiation damage – it is simply not safe to spend long periods outside, exposed to the vacuum of space. A lunar atmosphere would also prevent smaller asteroids from reaching the surface and turning your Moon villa into a pile of dust.

A solution for creating an atmosphere on the Moon was proposed by astronomer and science fiction writer Fred

Hoyle. The idea is quite simple to imagine but rather hard to realise. Simply change the orbits of some comets in the outer solar system and send them crashing into the Moon. Comets are around 50 percent water plus other elements that would form a natural atmosphere. Upon impact the comets would vapourise, dispersing these gases and water vapour to create an atmosphere. These impacts would also liberate water that is contained in the lunar regolith, which could eventually accumulate on the surface to form natural bodies of water.

The transfer of momentum from these comets could also get the Moon rotating more rapidly, speeding up its rotation so that it would no longer be tidally locked. A Moon that was sped up to rotate once on its axis every 24 hours would have a steady diurnal cycle, which would make colonisation and adapting to life on the Moon much more pleasant.

How many comets would be needed? All the ocean water on Earth could fit into a sphere of about 1000 kilometres in diameter. The surface area of the Moon is just over seven percent of that of the Earth. So a sphere of water around 400 kilometres across should suffice, which could be provided by a few hundred large comets. Impacting the Moon at 30 kilometres per second would be more than enough to cause the Moon to spin once in 24 hours.

Unfortunately, these ideas would fail to deliver a permanent atmosphere to the Moon. Over time the solar wind would blow away the newly formed atmosphere. In a thousand years the air would be gone, the temperature would drop and the water would freeze. But I have a solution for this: Let's place the Moon at the second Lagrange point of the

Earth-Sun system, where the magnetic field of the Earth would deflect the solar wind safely past our Moon. We already park satellites there that need to be in the permanent shadow of the Earth. This keeps them cool and allows us to measure the radiation from the big bang for example.

This special place in space lies behind our Earth and is in a constant eclipse of the Sun. An object placed there needs little energy to maintain its position – the gravitational forces from the Sun and Earth allow this object to move synchronously with the Earth. It is 1.5 million kilometres away, about four times the current distance to the Moon. The Earth's shadow at the distance of the Moon is 9200 kilometres, compared to the size of the Moon which is 3476 kilometres. But if the Moon were manoeuvred there then the Earth's shadow would cover just 40 percent of the lunar disk. From Earth, our Moon would look like a circle of light surrounding a glowing red interior.

Moving the Moon sounds like the stuff of science fiction, and it is, but good science fiction always has a chance of coming true. The comets used to create oceans of water and an atmosphere on the Moon could be placed on orbits that transfer energy between the Earth and the Moon. By sending the comets to orbit behind the Earth as it moves around the Sun, the comet steals some of the gravitational energy of the Earth and moves outwards, where it can be passed in front of the Moon, giving that energy to the Moon. The net effect after hundreds of orbits is to move the Moon outwards at the expense of moving the Earth closer to the Sun. But because the Earth is so much more massive than the Moon, the change in Earth's orbit is tiny compared to the change in the Moon's.

Once the Moon reaches four times its current distance it can be manoeuvred into the Lagrange point where it would seemingly hover, fixed in space against the night's sky. Small adjustments will be needed over time to maintain its position, since this is not a stable place to be, but that's easy to maintain compared to what we have achieved already. There the Moon will be shielded from the solar wind, the Earth's magnetosphere deflecting the particles safely past, allowing an atmosphere to be retained on the Moon. There would be some negative side effects though. At its new distance the Moon would no longer stabilise the obliquity of our planet – the tilt of the Earth would begin to vary and our climate would undergo catastrophic variations.

I hope this is never accomplished and our Moon is left as it is, to inspire future generations as it has past generations, to admire as an object of beauty in the night's sky and as testament to our solar system's grand and mysterious past. Unfortunately our Moon is going to end up out there anyway, regardless of whether we move it or not.

In order to understand the future of our Moon we can come back to the Moon's motion away from the Earth. Recall how the gravitational tidal interactions between the Moon and Earth cause the Earth's rotation to slow down and the Moon to drift away from the Earth. As the Moon moves away, its size in the night's sky slowly shrinks. The last total eclipse of the Sun will be witnessed from Earth in a little over 500 million years from now. After that time, the Moon will not appear large enough to block out all the light of the Sun.

In the far future the rotation of the Earth will have slowed down such that it turns once in the time the Moon takes to orbit the Earth – the system is fully tidally locked. That would happen tens of billions of years from now, and the Moon will have moved twice as far away and will take 50 days to orbit the Earth. The length of Earth's day will have slowed down to over a thousand hours long to match the orbital time of the Moon. Night and day on Earth will each be five hundred hours long, and the Moon will appear half its present size and will only be visible from one side of our planet. The Sun will be the main driver of the ocean tides, which would have decreased to about ten centimetres. But there will be no oceans at this time in the future. They will have evaporated in a few billion years from now as our evolving Sun heats up our planet far beyond what our human activities are capable of achieving.

Our star is about half way through its life, and in another seven billion years it will transform into a red giant star. Our dying Sun could engulf the Earth and Moon, turning them back into molten rocks as they were when they formed. But if they survive that inferno, then the small tides from the remnant of our star will continue to create a small tidal bulge on the Earth. The gravity of the Moon will pull on Earth's tidal bulge, speeding up Earth's rotation. As a consequence the Moon would continue to move back towards the Earth, until Earth's gravity tears the Moon into fragments, creating a ring of debris as spectacular as Saturn.

Since our species first gazed upon the world, the Moon has made around four million journeys around our planet. But in the eyes of our Moon, this is just a flicker in the passage of time. Our Moon will continue its cosmic dance, eternally interlocked by gravity with our world, partners from birth to the very end.

## ABOUT THE AUTHOR

Ben Moore is Professor of Astrophysics at the University of Zurich, Switzerland, where he is Director of the Centre for Theoretical Astrophysics and Cosmology. He has authored over 200 scientific papers ranging from the origins of planets and galaxies to dark matter and dark energy. A British citizen, he gained his PhD from the University of Durham and spent several years as a research fellow in the United States of America at UC Berkeley and the University of Washington. His research group simulates the universe using custom built supercomputers. Under his artist name "Professor Moore", he plays neoteric electro-rock music and has organised and played on a love-mobile during the Zurich Street Parade. Moore has been living in Switzerland for the past 17 years, where apart from his music he enjoys mountaineering, snowboarding and an occasional game of Quake.

Printed in Great Britain
by Amazon